ENVIRONMENTAL POLLUTION AND BHOPAL KILLINGS

ENVIRONMENTAL POLLUTION AND BHOPAL KILLINGS

Dr BROJENDRA NATH BANERJEE

GIAN PUBLISHING HOUSE
DELHI-110007

GIAN PUBLISHING HOUSE
29/6, Shakti Nagar, Delhi-110007

ISBN 81-212-0119-5

Printed in India
Published by Gian Publishing House, Delhi and
Printed at
Bharat Enterprises, Delhi-110053.

Lovingly Dedicated To

Raymond and Barbara Livingston
for their
fellowship and friendship

PREFACE

ENVIRONMENTAL Pollution did take place by the callousness of the US multinational Union Carbide Corporation in Bhopal. By and large, the prospects outlined are so awful that one catches oneself thinking that it is all too dreadful ever to happen. But the truth is that it has already happened. And the catastrophe occured where it was least expected—the Union Carbide Plant at Bhopal.

Why did it happen? What if any, are the medical and biological consequence of the accident? Those victims who returned to the area around the plant would be able to live a normal life again? Do they need special medical care and rehabilitation programme? Will such dangerous plants continue to operate in other parts of our country?

This gas catastrophe was a great misfortune to the victims. It involved human losses and massive material and moral damage. At the same time, this tragedy has brought to the fore the whole issue of the wisdom of continuing such dangerous industrial plants.

How should the calamity be assessed? These questions are being asked these days by the environmentalists. And assessments of what happened differ greatly. In the absence of a proper analysis of the tragedy and its consequences and without comprising the hazards posed by such industrial plants with those presented by other modern industries it was vital to enable experts to make a calm, objective and unbiased assessment of the Bhopal tragedy and its criminal killing. With this in mind, the author has made this honest attempt to find out the causes of the accident and its consequences on environment in a long term perspective. There is an effective solution to such merciless killing of innocent people around the plant. But, why did it fail?

Now, the victims should receive necessary assistance, including material help and employment in their new situations. After all the destruction, the lives of the victims have not been adequately sustained. Here is the tragedy of this environmental pollution as a result of discharge of poisonous MIC gas from the plant which passed into the environment. Should the need arise descendants of the victims may find a way of transporting it all elsewhere or of rendering it all absolutely safe. Such is the world we live in. Under the provisions of international law, individuals can sue for the compensation for damage in the appropriate courts, countless damage suits have been filed both at home and abroad (in the USA) within the course of three years. Immediately, no favourable judgments have been concluded to cheer the victims from the agony of their continual sufferings.

What we need today is the hectic efforts in dealing with the victims legally, morally and physically. The scope of such efforts has become necessary for the elimination of human sufferings so that mankind will be able to survive such environmental pollutions and degadation of resources. To day no task is more pressing and noble not only for an environmentalist but also for any sober-minded individual, than to prevent occurance of such insanity as Bhopal Gas Tragedy.

BROJENDRA NATH BANERJEE

September 1, 1987

C-17, Nizamuddin West
New Delhi-110013.

CONTENTS

1

BHOPAL GAS TRAGEDY—ANTOHER HIROSHIMA

UNION CARBIDE' DECEIT

UNION CARBIDE Corporation's (UCC) criminal irresponsibility ruined thousands of innocent lives as the killer gas walked out of their plant at Bhopal on December 2/3, 1984. The national tragedy at Bhopal was not a new one that was caused by the U.S. multinational Union Carbide. The shame of what happened on the night of December 2, 1984 rested squarely only on the shoulders of apathy and callousness of Union Carbide Corporation. All the roses of Arabia could not remove this ignominous stain of UCC in the recorded history of mankind.

December 2/3, 1984 was one of the coldest winter nights in Bhopal. The majority of more than 3.5 lakh people living in 30 out of 56 wards of the city were wrapped up in sleep. Suddenly, something wrenched the people from their slumber. They felt a burning sensation in eyes, then fits of choking and breathlessness: "As if a godown of red chillies had been caught fire", was how many a survivor described the unprecedented disaster experience of that dreadful night. This conjection caused the death of several thousands who might have been alive even today just like us, only they knew what it was they were inhaling and made an escape for life.

Thousands died because they did not know what catastrophe had hit them completely unaware while they fast asleep inside the warmth of the blanket or quilt. Unfortunately, they ran in the direction of wind carrying the lethal Methyl-Isocyanate

(MIC) gas and succumbed to life. They died like flies. Every gas victim had the Hobson's choice. Sheer chances decided who was to survive and who was to perish; who was to escape the death trap of the killer gas and who was to continue to perennial sickness in life. Thus another Hiroshima was repeated in human history by the US multinational corporation.

It was evident that besides MIC, other potentially more hazardous gases escaped from the tank numbered 610 inside the Bhopal plant. These included hydrogen cyanide gas, carbon monoxides, and a host of other more deadly cynogen chlorides. At least three distinct chemical reactions could be postulated from the three clear sedimentation levels etched on inside the tank, and that the temperature reached 1300°C, at which point the tank's stainless steel metal became melleable (the melting point of steel was 1800°C). The deaths were not caused due to MIC intoxication alone. According to the forensic department, hydrogen cyanide, and probably other more dangerous cynogen chloride. The latter group of poisons was particularly vicious for living organisms in even cold weather. The earliest trace of cyanide like poisoning were identified by four toxicologists led by Prof. Heeresh Chandra, director of the Medico-Legal Institute, toxicologist and Medico-Legal adviser to the Madhya Pradesh (MP) Government who was the first to identify the cause of deaths and confirmed by Dr. Sriramachari of the Indian Council of Medical Research (ICMR), Dr. Max Daunderer, head of the Munich-based Toxicology Institute and author of a several volumed work on toxicology who came well equipped with NATS ampoules and another CBI (Central Bureau of Investigation) toxicologist.

So the killer gas at Bhopal plant of Union Carbide Corporation was a cloud of toxic gases rather than mere MIC being released on the fateful night of December 2/3, 1984. The dean of the medical college, Dr. BBL Mathur, in a written record of a meeting with Dr. Daunderer, stated:

"This NATS treatment was given to some patients in paediatric wards who showed overall improvement within hours." Dr. Mathur carefully listed the various stages of gas intoxication:

"Methyl-isocyanate has two compounds, a local and the systematic irritant—the stages of intoxication—

(i) eye, skin and throat irritation,

(ii) after hours, days and weeks lung oedema,

(iii) after re-absorption and metabolism cyanate intoxication,

(iv) after acute cyanide intoxication possible anoxia of all organs,

(v) long term brain disasters."

The gases generated a disaster because they affected living organisms on a massive scale. They were toxic and harmful of natural processes. Let loose in a natural eco-system, they disrupted ordinary biological processes, causing mortality in some and severe long term disability in thousands of others. That was a simple and inevitable consequences. In the case of the Bhopal gas disaster, the most conservative estimates put the number of people dead as 2,500 (according to non-official estimates the death toll rose to more than 15,000); in addition, more than 2.5 lakhs were victims of disabilities, some permanent. All gas victims during those few hours of the leak had, in addition to the medical disabilities inflicted on them, to undergo a fearsome kind of terror and mental anguish. In American court such an experience of an unparallel catastrophe by itself, would have got these hapless victims a compensation no less than $ one million per family.

Thousands of survivors in the acute phase suffered from severe symptoms referable to respiratory, gastro-intestinal, ophthalmic and musculoskeletal systems. In addition, the psychic trauma caused by the tragedy with the resultant feelings of anxiety, insecurity and depression only made the situation worse. This not only generated new symptoms of varying nature but also intensified those due to organic damage. Eyes of about 200,000 individuals were affected in varying degree depending on the duration of exposure. There were still widespread complaints regarding watering of the eyes and "foreign body" sensation.

Symptoms of gastro-intestinal cases were experiencing lack of appetite, burning sensation, etc. Endoscopic examination of such patients revelaed the presence of gastritis, oesophagitis and ulcers. The traumatic event left in its wake a large number of people—men, women and children—who suffered from serious anxiety and other forms of mental distress and showed signs of psychiatric morbidity. The survivors in the acute phase showed symptoms such as anxiety, panic, depression, feelings of insecurity, fear of death, helplessness, disturbed sleep.

Till the end of October 1985 studies conducted in 393 families with a total population of 1195 revealed psychiatric morbidity in 193 subjects. The majority of them were found to be suffering from neurotic group disorders like neurotic depression (66.3 per cent) and anxiety state (30.3 per cent). Out of 855 patients in 10 government-run clinics, 193 (22.6 per cent) were found to be suffering from mental disorders. Of these 37.3 per cent were suffering from neurotic depression, 24.9 per cent from anxiety state and 35.2 per cent from adjustment reaction. Till the end of October, 1985, 573 patients attended the psychiatric clinic in Chola Road Polyclinic. The majority of them were suffering from neurosis depression (41.7 per cent) and anxiety state (26.4 per cent), 9.6 per cent suffered from psychoses. Patients at these clinics were managed by anxiolytic, anti-depressant and anti-psychotic medications as well as by psycho-therapy. Many people abandoned their homes and left Bhopal, losing their livelihood and incurring heavy expenditure on travelling. For the people at the lowest rung of the economic ladder thus expected a heavy toll. For nearly two months all trade and commerce, industrial activity and development work came to a near stand-still situation. There were thousands who were unable to do any work. Thousands of others were tending their own sick relations.

The traumatic events of December 2/3, 1984, following the tragedy left in its wake a large number of men, women and children who suffered from serious anxiety and other forms of mental distress. In fact, they showed signs of psychiatric morbidity. Post-disaster syndromes are observed in survivors of floods, earth-quakes, fires and other calamities. The Bhopal industrial disaster was different from other calamities because, apart from causing acute mental distress to the survivors, it also left them physically ill due to the toxic effect of the gas which leaked from the Union Carbide plant.

The extent and nature of psychiatric problems are being investigated by a mental health team comprising psychiatrists, and social workers under the guidance of Prof. B.B. Sethi, principal, King George's Medical College, Lucknow, and his associates in callaboration with Gandhi Medical College, Bhopal and the Bopal district health-authorities. An initial pilot survey in gas-affected areas revealed that the gas tragedy had resulted in mental disorders on a fairly wide scale. The team under the sponsorship of the Indian Council of Medical Research (ICMR) has been conducting a detailed general population survey of the gas-affected region since June, 1985.

In a larger context, the fall-out of the Bhopal gas tergedy has been another out-cry around the world against the profit-at-any-cost attitude and the callous cost-cutting practices of the US—multinationals in the developing countries. Such practices, played on unwary governments largely innocent of the dangers lurking in several production processes have wrought unimaginable disasters, which are scarcely noticed. Unless a Bhopal or Seveso hits the public consciousness. In the U.S. multinational Union Carbide plant in Beziers in Southern France, Liquid MIC is kept in stainless steel drums in a specially built concrete warehouse. There is a computerised system to detect the gas in air in fraction as small as 0.3 parts per million (PPM).

MIC is itself a pesticide, and, according to handbook in pesticides manufacturing, is indeed used to kill nemadotes. However, its more common use is as an "intermediate." a chemical that is used in the last or penultimate stage of pesticide or drug manufacture. At the Union Carbide plant, MIC is reacted with another chemical, called alpha-napthol, to yield the pesticide, carboryl. The US-UCC multinational has specialised in using MIC as an intermediate for its other pesticides too.

MIC is a fairly stable liquid at room temperature, except during summer. It has a boiling point of 39.1 degree Celcius. There are two problems with MIC. One is storage. MIC reacts vigorously with water (including moisture), alcoloids and aminoes as well as many common solvents. Hence it must be stored in completely inert condition and kept cool. This is an expansive and risky operation. That is why the storage and use of MIC is severely restricted in many countries.

The second problem is its toxicity. MIC can kill in very small doses, when ingested or inhaled and even when absorbed through the skin. This last factor makes it an even greater hazard from say, phosgene, which affects humans through the inhalation route.

Handbooks on industrial hazards and poisonous chemicals rate MIC extremely toxic. The threshold limit value (TLV) on the maximum permissible exposure limit for industrial workers is only one-fifteith of one part per million for MIC. For hydrogen cyanide it is 10 parts per million and for phosgene one-tenth of a million.

MIC is highly irrritating to the skin, eyes and mucuous membrane—which is one of the main causes for the early deaths in Bhopal. It is difficult to get an unambiguous account of MIC's smell. But it is unpleasantly pungent and yet sweetish by most accounts.

MIC is not as well understood as many other chemical poisons since it is much less commonly used. No antidote to it is firmly known. However, laboratory experiments on animals show that MIC can kill them in minute doses. A rat that breathes air containing only five parts per million of MIC for a mere four hours has a 50 per cent chance of dying. Its effects on human beings have not been adequately studied for obvious reasons. What is not known, in particular, is that MIC is like many other toxic agro-chemicals, as a cancer-causing and metagenic substance. Its long-term effects in low doses are also not known yet. But they are suspected to be extremely lethal.

Methyl Isocynate (MIC), a product that few in Bhopal knew even existed, which Union Carbide, its manufacturer, described in their manuals as "an extremely hazardous chemical," a poison against "stringent pecautions must be observed to eliminate any possibility of human contact." Yet it was the lethal substance, a highly volatile chemical that was used by Union Carbide to manufacture pesticides, that vapourised silently into Bhopal and converted a pleasant mild winter of the 3rd December night of 1984 into a nightmare of misery, panic, sickness and death for adults, children, fathers and mothers, siblings and grand-children a slow painful and unnecessary snuffing out of life.

The US multinational Union Carbide had to face a social and legal indicment which no other multinational or national company was confronted in India before the Bhopal gas tragedy.

UNION CARBIDE COMPANY

The 1984 annual report of Union Carbide India Limited (UCIL) displayed half a dozen smiling faces on the cover with a discreet blurb in one corner, "serving the community," and "touching the lives of many people in many ways." Sales in 1984 were Rs. 212.54 crores, with pre-tax profits of Rs. 14.77 crores. In terms of sales, that put the company, which operated plants in nine cities, among the top 20 Indian companies. The Union Carbide factory came to India as EverReady company and was incorporated in Calcutta under the Indian Companies Act on June 21, 1934. In 1941, the name of the company was changed to National Carbon Company. In 1956 it was again re-named as Union Carbide India Ltd (UCIL). In May 1969 the company set up a modest formulation plant on a five-acre plot at Kali Parade ground in Bhopal. The letter of intent was signed in 1972. Three years later, in 1975, the Indian government granted Union Carbide a licence to manufacture pesticides. The company was set up with 51 per cent majority ownership by Union Carbide Corporation of the USA, with the other 49 per cent in the hands of private Indian companies and individuals.

The part of its operation currently the centre of so much unwanted attention, the Agricultural Products Division was started in a small way in 1966-67 when an office was opened in Bombay. In 1968 the division was shifted to Bhopal and work started in the formulation plant: the sevin technical concentrate was imported from the USA and only the work of blending and grinding was done here. The next year process development was started to make Temik, a highly-toxic pesticides which are also MIC-based. MIC was to be imported from the USA for many years to come and in 1973 the first MIC drum (181.8 k.g.) was brought in from the USA for experimentation.

Subsequenly, the Bhopal plant spread over an area of about 90 acres of land when it started manufacturing methyl-isocyanate based pesticides with the brand name Sevin. In fact, methly-isocyante is an intermediate product in the manufacture of sevin. Methylisocyante is produced by reacting phosgene with monomethylamine. Phosgene is produced by the reaction of carbon monoxide with chlorine.

$$2C+O_2 = 2CO$$

Caloinated code code oxygen — Carbon monoxide

$$CO=CL_2 = COC\ I_2$$

Carbon Monoxide Chlorine — Phosgene

$$COC\ I_2=CH_3\ NH_2 = CH_3\ NCO=2HCl$$

Phosgene Mono-methylamine — MIC Hydrochloric Acid

$$CH_3\ NCO=C_{10}H_7\ OH = C_{10}\ H_7\ CONHCH_3$$

MIC L-Naphthol — Carbaryl or SEVIN

MIC was being stored in three steel containers each with a capacity of approximately 60 tonnes. MIC is a toxic gas and is unstable. It is supposed to react violently with water and rust and possibly can react with itself under ceratin conditions to form polymers. All these recations are exothermic in the sense that the reaction produces heat. The boiling point of MIC is 39°C and it is a heavy substance with specific gravity of 0.96 (specific gravity of water is 1). The manufacture of MIC had stopped in the Bhopal plant sometimes in October, 1984 and the MIC plant was shut down for maintenance. MIC was stored in two tanks and one tank was supposed to be empty. The tanks were numbered 610, 611 and 619.

It was from the tank No 610, which, it is estimated, contained about 43 tonnes of MIC that the leakage took place. Why did the gas leak was known only after various investigative agencies and the Enquiry Commission completed their investigations. What was the composition of the gas which leaked from the plant? It was understood that MIC started dissociating at temperatures beyond 250°C. The MIC was dissociated into hydrogen cyanide and carbon monoxide. The lethal gas which leaked at the Bhopal plant was a mixture of MIC and other killer gases.

A year earlier, the company had received a letter of intent for the manufacture of 5,000 tonnes of MIC-based pesticides: the first phase of the project was completed in 1977 and production was started, the second phase of construction being completed towards the end of 1979.

In 1977 the Bhopal plant produced 321 tonnes of MIC-based pesticides. The MIC was imported in special stainless steel drums, and both the Sevin and Temik varieties used it. In the first half of the '70s sales rose from about Rs. 3 crore in 1971 to Rs. 16 crore in 1976 and further to over Rs. 20 crore in 1979-80—they were down to half that figure. The downslide began in 1977-78 when a consignment of MIC was delayed at sea. Coupled with drought and the consequent poor purchasing capacity of the farmer, this led to a fall in sales and by this time other pesticide manufacturers were also in the fray. The division never found its feet after that and sales sank after 1980. Ironically, it was in February 1980 that the MIC plant went into roduction.

Indeed, the US multinational Carbide story read like an epic of corruption-cum-nepotism all the way down from Janata government, which imported Carbide's used and second-hand gas chamber into Bhopal, to the Congress ministers and bureaucrats both at the state and the central government levels who permitted it subsequent rampage in Bhopal.

The Bhopal plant's location was controversial from the beginning. As early as 1975 M.N. Buch, a local official in charge of public services, tried to shift the location, but nothing happened. Instead Buch, himself, was removed to another post soon after Union Carbide donated Rs. 25,000 to the city municipality for a park. There were strong ties between Union Carbide and some local politicians. One well-known Congress (I) member became its lawyer. A former police chief of Madhya Pradesh was employed as security adviser after his retirement. A nephew of the state's former education minister was hired to handle public relations. And the brother-in-law of a deputy chief secretary of M.P., took on a high level job at the factory. Lavish parties for local dignitaries were often thrown at Carbide's luxurious guest house in the Shyamala Hills. In 1983, when Congress (I) held a regional conference in Bhopal, VIPs were put up at the guest house. The country had to pay heavily for all this on the night of December 2/3, 1984.

Nothing illustrated the strange and weird collusion between administrators, scientists and doctors better than the question of cyanide and its role in the Bhopal gas tragedy. The manner in which officialdom, the establishment, clamped down in one way or the other on any suggestion or clinical evidence that the victims had died from cyanide poisoning or that those surving continued to suffer from a form of chronic cyanide poisoning, was inexplicable unless either one or both of the following were true.[1]

First, that those involved in the cover-up, though occupying high positions in the medical and official establishment, were due to some events in the past, beholden to the US multinational corporation (MNC). The leading representative of this group was Prof. N.P. Mishra whose department of medicine, Hamidia Hospital, according to UCC's own testimony before the Bhopal Poisonous Gas Leakage (1984) Enquiry Commission, had been gifted four expensive sets of equipment, and a private ward consisting of two suites constructed by the company.

To distract attention from cyanide poisoning, the Hamidia hospital establishment put forth and defended, a theory of MIC damage that was limited to pulmonary damage and dysfunction in a small proportion of the population.

---

1. The Illustrated Weekly of India, December 1, 1985.

Second, that the state and central governments could simply not acknowledge cyanide poison due to its implications; it was beyond the capacity of the administration to care for such a large proportion of physically and mentally ill, most irreversible so, and with steadily deteriorating health, and therefore, better for it to restrict damage publicly to one or two major symptomatic disturbances, and rule everthing else, as of no medical significance.

The suggestion that there was no long term damage among the victims, first broached by the Wall Street Journal had, therefore, practical implications not only for the Union Carbide (in terms of liability) but also for the central and state governments in terms of long term relief and rehabilitation.

Dr. Heeresh Chandra, director of the Medico-Legal Institute, Toxocologist and medico-Legal adviser to the M.P. government, was the first person to identify the cause of deaths on that December morning as due to cyanide poisoning. On December 3, as he cut up one corpse after another, Dr. Heeresh Chandra was overwhelmed. The signs were unmistakeable: the cherry red colour of the blood, lack of cynosis on the face, froth in mouth and nose, hypostasis red, not only of the dependable parts, but all over. Organs showed haemorrhages, particularly the lungs, externally as well as internally. All these signs constituted evidence of acute phase cyanide poisoning, among others. Dr. Chandra would postulate his conclusion that would become almost historic in the days ahead:

> "Death is due to poisoning of the irrespirable gases resulting in cytoxic or hystoxic anoxia".

The antidotes for cyanide poisoning are sodium thiosulphate and sodium nitrite. Sodium nitrite is toxic and is, like snake anti-venom, to be administered only if there is positive proof of toxins in the body. It was not available in Bhopal, whereas sodium thiosulphate (NaTS), non-toxic, was available in plenty: known as hypo, it is the basical chemical used in all photo labs.

On December 4, the head toxicologist of the CBI joined Dr. Chandra at the autopsies. Unknown to these forensic experts, a telex message had also come in on December 3, 1984 from

Dr. Bipin Avasia, medical director and toxicologist at the Union Carbide Institute plant in the USA, which included five points of treatment as an emergency measure. The third one read:

> "If cyanide poisoning is suspected, use amyl nitrite. If no effect, sodium nitrite and sodium thiosulphate."

Well versed in MIC technology, Carbide had already decided that cyanide had been liberated with MIC from the tank. The telex message, however, was a bad slip on the part of the US multinational corporation.

A study by the Economic Times Research Bureau revealed that dividend remittances by Union Carbide India Ltd over the years to the US company surpassed the aggregate investment made by UCIL. The Bhopal plant was transferred second hand from the MIC plant at Danburg, USA by Union Carbide Corporation. A change in the Bhopal plant design was one of the chief reasons for the MIC leak. According to sources, the CBI had evidence, the change in design was executed on the directions of the company's headquarters in the USA. MIC leak was reported at the Carbide plant at West Virginia, USA. This was, however, not the first leak. In the past five years, according to the company's own figures, there were 190 small gas leaks at the plant and 61 of these leakages were of MIC, 107 of phosgene and remaining 22 of combined MIC and phosgene.

A study by the Delhi Science Forum revealed that production of pesticides in the Carbide plant was, perhaps, only a cover for the multinational's vast research and development activities connected with chemical and biological warfare. On the other hand, according to Dr. Philippe Rhouda, a French specialist on medical emergencies, India's gas tragedy attracted chemical warfare experts interested in studying the killing efficiency of MIC. In memory of the thousands dead in Bhopal due to the machinations of the multinational Union Carbide company, several organisations held functions to commemorate the dead. The All India Peace and Solidarity Organisation (AIPSO) was one such organisation.

It also pointed out that Bhopal must become the symbol to combat the menace of the US multinational corporations (MNCs). It also organised a national conference on the role of multinational corporations in spreading the arms race, manipulating the media and subverting the sovereignty of nations. All this has been perpetuated so that its high profits would not be jeopardised.

Bhopal had become synonymous with not only a genocide being the ghastly scene of the biggest industrial mass murder, but also the most telling living testimony of the anti-people operations of the MNCs like the UCC. Bhopal once more proved that the multinationals have no frontiers, that they owed no commitments other than making profits. The multinationals were the most powerful instrument of imperialism through their sophisticated and intricately planned operations which included overthrow of even legitimate governments in the developing countries.

In the name of transfer of technology, many dangerous processes were shifted to the developing countries by the multinationals from the West. The claim sometimes made by certain interested circles, especially within the government and monopolies that MNCs assist developing countries in their growth through transfer of technology was a myth. Bhopal in any language today meant the depredation of the MNCs. The New International Economic Order (NIEO) which could strengthen self-reliance in development was a serious challenge to the MNCs. Therefore, it had been repeatedly thwarted particularly by the US administration. Every effort should be made to discourage the entry of MNCs in the developing countries. Wherever the MNCs operated now, governments must impose strict control over all their activities. The need for vigilence was further hightened as the Research and Development (R and D) activities of many MNCs, particularly in developing countries, were being carried out using subterfuges and clandestine methods for biological and chemical warfare systems.

All the developing countries had the common experience of their life styles, thought-process and cultural traditions being eroded and subverted by the virtual monopoly control enjoyed by the MNCs over the mass media. Massive mis-information and dis-in formation flooded the media of mass communication, especially the electronic media. This was highlighted by McBride Commission organised by the UNESCO which brought forth the wrath of countries like the USA and the U.K.[1] It was not enough to formulate a code of conduct for the MNCs merely as guidelines for their operations in various countries. The code must be mandatory not to with-hold any information about their activities.

There must be an end to the manufacturing of hazardous and toxic chemicals in developing countries since the chemical industry was now being militarised more. People everywhere should be urged to establish zone free from chemical warfare agents, since the lives of citizens in developing countries held very little value in the eyes of the Pentagon and the multinationals from the western countries. Union Carbide was the company that wilfully chose a hazardous techonology without any regard for safety, that wilfully neglected vital safety measures, that was responsible for gross under-designing of crucial safety systems, as also for completely inadequate instrumentation and control systems in the plant. The same corporation was responsible for deploying untrained operators and supervisors in critical areas to cut down costs.

Union Carbide provided no public information regarding the danger to public safety from the plant. It conducted research in the grey areas between chemical and biological warfare, and to this day, it withheld desperately needed information on the toxic nature of MIC, thereby greatly increasing the death and casualty toll.

According to Mr. Warren Anderson, president of Union Carbide Corporation, "people are safer in a chemical plant than in their own home." A PTI (Press Trust of India) report revealed that Carbide's research centre in Bhopal had been engaged in the controversial study of synthesising and field testing new chemicals for the last six years, right under the nose

1. Brojendranath Banerjee: Concern to New Delhi. South-South Corporation. Paribus Publishers and Distributors, New Delhi (1983).

of the department of science and technology (DST) in the garb of agricultural research. Moreover, the centre had been sending germplasm to Union Carbide in the USA, which entered the business of seed-breeding in a big way. No wonder, the parent company had pumped in more money for its research centre in Bhopal while its pesticides plant, 10 kms away was suffering huge losses since 1980. Unchastened by the ghastly tragedy and dismissing reports of its R & D centre being involved in research on biological warfare as "baseless," Union Carbide sought government's fresh sanction to produce two killer gases —phosgene and MIC—this time for merchandising purposes as also exports. Meanwhile, the mystery surrounding the company's R & D centre deepened further as a group of visiting newsmen were refused entry to the sophisticated Union Carbide's chemical laboratory located on the Rs. 40 crore Agricultural Products-Division (APD) of the Union Carbide in Bhopal.

LICENSED TO KILL

The killer gas escaped into the air from the leaking Union Carbide plant and suddenly enveloped the Old Bhopal city killing thousands and maiming many more. Bhopal today was no longer the centre of speculation and debates regarding the cause behind the world's worst industrial disaster. While the parent Union Carbide Corporation (UCC), USA obstinately was clinging on to its sabotage theory as the reason behind the tragedy, investigating angencies in India pinpointed on the legal and moral responsibility on the US multinational and urged for maximum compensation for the MIC-affected victims. Union Carbide from the outset disowned responsibility for the accident, hinting at the fact that the Indian management was liable for manufacturing safety standards at the plant. However, documents confiscated by the CBI included administrtaive manuals of the company which categorically stated that any of their divisions with less than a 50 per cent holding was their subsidiary. UCC in official records, now with the CBI, was shown to have a 50.9 per cent holding in its Indian counter part. The MIC unit at Bhopal had shown signs of wear and tear much before the catastrophe and a bit more concern and vigilance on the part of the Indian managers would have saved many lives.

There were other documents to prove that not only did the parent company execute financial control over the Bhopal plant, it also exercised administrative jurisdiction over the APD division to the extent of appointing staff, making transfers and minor investments. Confidential files collected from the Bombay headquarters on September 27, 1985 had manuals of international procedures circulated by the UCC which clearly described the Bhopal plant as its subsidiary. There was every indication that the safety standards at the Union Carbide Bhopal plant were not as stringent as the UCC's MIC plant at West Virginia, USA. Absence of sophisticated security systems at the Bhopal unit caused the collapse of the in-built three-tier security system, which should have been activated in case of even a minor leak.

The inoperation of these systems or their defects was the proof of the negligence of both the Indian officials and their American bosses, since it was known to them that in case of a thermal decomposition, MIC could polymerise into cyanide. The data sheet on methyl isocyanate recovered from the Union Carbide premises (No F 43458A) stated in the section marked reactivity data:

> "Thermal decomposition may produce hydrogen cyanide, nitrogen oxides, carbon monoxide and carbon dioxide."

Knowing fully well the dangers of strong MIC in such large quantities (tank No 610 had a capacity to store 90,000 pounds of MIC), the UCC had no excuse to do away with the latest safety devices in Bhopal which they had adopted in their West Virginia plant. Unlike the American MIC unit where there was a claim to have an auto-warning alarm system, at Bhopal, a leak could only be monitored manually. Besides, examination of the chemical reaction which triggered off the leak shoed that whatever safety methods had been adopted by Union Carbide were rendered inadequate in the face of the catastrophe.

A valuable piece of evidence against Union Carbide was the dossier of proposals sent by the UCC to the Bhopal management regarding the shifting of the MIC units to Brazil. The Bhopal MIC unit was a losing concern and with losses mounting, plans had been finalised for shifting the entire plant lock, stock and barrel out of India. A letter from R. Natarajan, the vice-president of UCC, Hong Kong, to K.S. Kanda, senior official UCIL, Bombay, a copy of which was sent to V.P. Gokhale, stated that a MIC unit was to be installed in Brazil. The letter added:

> "There is also a possibility that a carbomolation facility will be installed in Indonesia. The question is whether economically the MIC facility and the caramolation facility in Bhopal could be dismantled for shipment to these countries. . . Since the labour cost in India was low, such a move could be effected in a viable manner."

In reply to the proposal, a letter was sent to the Hong Kong division which was important. It stated:

> "The shipment of the MIC unit is a problem because of the high corrosion at several points and some of the tall columns."

The letter clearly indicated that the condition of the MIC plant was far from satisfactory. After this reply, detailed financial proposals of shifting the MIC unit to Brazil were to be telexed to Hong Kong but the tragedy prevented such a move. Four days before the mishap, on November 29, 1984 these proposals were to have been sent from Bhopal and were among the documents recovered from the office premises when the CBI took charge of the Union Carbide plant. This, according to CBI officials, was the most incriminating evidence against UCC. It established that UCC was not interested in the MIC plant any longer.

The log books maintained by the different departments of Union Carbide indicated that the pressure inside tanks 610 and

611 containing methyl-isocyanate (MIC) was frequently increasing and posing concern. But no serious efforts were made to rectify the fault or stop production. Instead, the Carbide management chose a short cut: the tanks were depressurised by releasing small quantities of the gas into the atmosphere through the flare tower. Moreover, when the management wanted to transfer the MIC from the tanks to the Sevin unit and found that the gas was not responding to the prescribed nitrogen pressuring method, it chose to pressurise the tanks by a different method. An examination of Log Books left no doubt about the criminal negligence on the part of the Bhopal management.

Extracts from different log books are being reproduced hereunder:[1]

Production supervisor's log book

November 15, 1984, shift 1 Tank 611 to be depressurised after starting flare tower, level tapping 611 (HP) is plugged. Such an entry indicates that the flare tower was not in working order.

Production supevisor's log book

November 15, 1984, Shift 2 Tank 611 is isolated as pressure is increasing.

Production supervisor's log book

November 16, 1984, shift 1 MIC tank 610 vent valve is isolated. shift 2. MIC tank 611 depressurised upto 1.5 kgs.

November 24, 1984. 1,175 litres of MIC transferred to Sevin unit after arresting supply line (flange) leak. The entries in the production supervisor's log book on November 25, 1984, indicate that the situation was normal on all the shifts. But one entry regarding tank 611 reads:

"Suspected dip tube choked. Tapping found isolated. To be followed up."

1. SUNDAY, 14 December, 1985.

However, further entries do not reveal whether any follow-up action was taken or not. Standard operation procedures were also violated during the unloading of chlorine as a result of which the gas leaked into the nitrogen header and contaminated the entire nitrogen supply system. Incidentally, the contaminated nitrogen entered tank 610 and 611. The following entries in the production supervisor's log book should serve as an eye opener.

November 28, 1984. Shift 1: There was a minor chlorine gas release while unloading Chlorine backed into nitrogen header because nitrogen valve was not closed properly.

Shift 2: Nitrogen header depressed into vent gas scrubber (VGS).

Shift 3: Nitrogen header purged out. Still slight chlorine smell persists.

November 29, 1984. Shift 1: Phosgene tank pressure safety valve dipping checked for phosgene/MIC. No traces found. Main valve on RVVH (Relief Valve Vent Header) closed. All filters isolated and bleeders kept open.

The entries on November 30, 1984, reveal the extent of the danger posed by the MIC unit. Both the MIC storage tanks 610 and 611 were behaving erratically. There was a desperate attempt on the part of the management to generate pressure inside the MIC tanks so that a portion of it could be transferred to the Sevin unit. Standard operation procedures were violated and tank 611 was pressurised with the help of a copper tubing despite the fact that copper reacts adversely with MIC. The following are relevant excerpts from the log books:

Production supervisor's log book:

> November 30, 1984. Ref. tank number 610. Nitrogen is going into tank but no rise in pressure.
>
> Plant supervisor's log book: MIC tanks not getting pressurised, so MIC not flowing to Sevin.

Production supervisor's log book:

Shift 1: Tank not getting pressurised. No make up. N2 is coming. At shift change isolation valve on N2 line kept closed.

Shifts 2 and 3: pressurised through Pressure Indicator (PI) Bleeder and transferred 2,150 kgs. of MIC to Sevin. MIC tank DMC found stuck in closed position. DMV opened from bonnet, cleaned and fixed. DMV operating well. N2 line of tank 611 strainer plug opened. Its D/S line is choked. Line disconnected.

Further entries in the log book on subsequent days reveal that the pipes and valves leading to tank 610 and 611 were choked due to the solidifying of MIC in the presence of water. Instead of investigating the causes behind the presence of water, the operating crew cleaned the pipes and valves and continued the process of transferring MIC from the storage tanks to the Sevin unit.

On the fateful night of December 2, 1984, the log book recorded the leak shortly past midnight.

Sevin unit log book: Shift 3: Operation suspended due to heavy MIC leak from plant at 12.45 a.m.

On December 3, 1984, when Bhopal was reeling under effect of the poisonous gas, when hundreds of people and cattle lay dead on the streets, the log book of the instrumentation department read:

> "Vent gas scrubber flow and level were reported faulty. Purged with water. LIC of VGS one plunger was arranged from shop and fixed. Caustic withdrawn from field stage 8.14 tonnes (100 per cent basis)."

PERFORMANCE OF MULTINATIONAL IN INDIA

There are more than 2,500 public limited companies in India, which are listed on one or more of a dozen stock exchanges spread throughout the country. However, the top 500 companies (Aridhi's 500) account for 98 per cent of trading activity. The market capitalization of Aridhi's 500 in early September, 1985 was Rs. 16,400 crores (US$ 13 billion). One fifth of Aridhi's 500 are subsidiaries of multinationals and the remaining four-fifths are India companies. An average multinational subsidiary in India has sales of Rs. 75 crores ($60

The MIC storage tank No. 610 being dug up for examination

In the presence of a special team from the parent Union Carbide Corporation (UCC), USA, tank No. 610, from which the deathly gas had leaked, was dug up, cut open and samples of MIC residue handed over to the UCC officials and CSIR scientists.

million) on an employed capital of Rs. 30 crores ($24 million). The resultant capital turnover ratio is 2:9. Equity capital constitutes nearly 60 per cent of capital employed, and consequently, the debt equity ration at 0.7 is rather low.

Multinationals in India have operated with a very low gearing, perhaps because so far they have not felt the need to borrow aggressively. This has resulted from a combination of two factors—various controls which the government has started relaxing only recently, and the hesitancy of multinationals to expand in an uncertain environment. Indian companies are relatively larger in size. They also have a higher debt equity ration of 1:8 and a lower capital turnover ratio of 1:7. These figures reflect their concentration in capital intensive areas with a much higher dependence on borrowed funds. The net profitability of multinational companies is better than that of Indian companies. The difference, however, narrows down if the high interest and depreciation burden of Indian companies (partly offset by lower tax rates) is discounted.

Typically, a multinational earns 16 per cent of sales as gross profit before depreciation, interest and tax. About 41 per cent of gross earnings are taken away by interest and depreciation charges. Of the remaining amount, 46 per cent is paid to the exchequer as tax. Overall, net profit amounts to nearly 20 per cent of net worth. Gross earnings of Indian companies at 17.5 per cent of sales are slightly better. Interest, and depreciation are much higher. As these are charged to expenses, the effective tax rate at 15 per cent is much lower. The net profitability of top 100 Indian companies, as measured by profit after tax to net worth, is 15.4 per cent compared to 19.6 per cent for top 100 multinational subsidiaries in India. The top 100 multinationals played an average dividend of 20 per cent on equity. One-fourths of the multinational group paid a dividend of 25 per cent or more, the highest being 110 per cent. Indian companies, by comparison, paid a much lower dividend of 9.5 per cent, and the average for even the top 100 Indian companies was only 16.7 per cent. There were only 3 Indian companies in Aridhi 500 which paid a dividend of more than 25 per cent.

The higher dividend is reflected in higher yield which averaged 3.2 per cent for multinationals against 2.6 per cent for top 100 Indian companies. In stock markets, scrips of multinational subsidiaries command premium. The price earnings ratio of multinationals is 15.3 against 10.6 for remaining 80 per cent Indian companies in Aridhi 500. Variations in depreciation and accounting practices in India distort the significance of price-earnings multiple. Aridhi has found that price-gross profit is a better multiple to compare market performance. Equity shares of multinationals are selling 4.9 times the gross profit per share against a multiple of 3.0 for remaining Indian companies in Aridhi 500. The stock market performance of multinationals has been distinctly superior. Appreciation in market price during the three years 1982 to 1984 averaged 48 per cent per annum for top 100 multinational against only 25 per cent for top 100 Indian companies.

Comparative performance of Indian companies

S.	*Indicator of Performance*	*Top Hundred*	
No.	*Multinationals*	*Indian Cos.*	
1.	Sales to Capital employed (Ratio)	2.9	1.7
2.	Debt to Equity (Ratio)	0.7	1.8
3.	Profit before Tax and Interest to capital employed (%)	35.9	17.5
4.	Gross profit to Sales (%)	15.9	17.5
5.	Interest to Gross Profit (%)	22.3	36 0
6.	Depreciation to Gross Profit (%)	18.5	28.0
7.	Profit Before Tax to Sales (%)	9.4	4.6
8.	Tax rate (%)	46.1	15.2
9.	Profit after Tax to Net Worth (%)	19.6	15.4
10.	Dividend (%)	20 0	16.7
11.	Yield on market price (%)	3.2	2.6
12.	Dividend Payout (%)	50.0	35.3

Size of Indian companies: small by US standards

Top Hundred

Indicator of Size	*Multinationals*		*Indian Companies*	
	Rs. Crores	*US$ billion*	*Rs. Crores*	*US$ billion*
Market Capitalization	4360	3.5	8975	7.2
Sales	7805	6.2	15915	12.7
Capital Employed	2995	2.4	10030	8.0
Profit After Tax	225	0.2	720	0.6

Source: The Times of India Jan. 7, 1986.

FESTIVAL OF DEATHS BY THE UNION CARBIDE

The US multinational plant was producing a pesticide with the brand name, Sevin. Methyl isocyanate is an intermediate product in the manufacture of Sevin and it is produced by reacting phosgene with monomethylamine phosgene which is produced by the reaction of carbon monoxide with chlorine. MIC was being stored in three steel containers each with a capacity of approximately 60 tonnes. MIC is a toxic gas and is unstable. It is supposed to react violently with water and rust and possibly can react with itself under certain conditions to form polymers. All these reactions are exothermic in the sense that the reaction produces heat. The boiling point of MIC is 39°C and it is a heavy substance with specific gravity of 0.96 (specific gravity of water is 1). The manufacture of MIC had stopped in the Bhopal plant sometime in October, 1984 and the MIC plant was shut down for maintenance. The MIC was stored in two tanks while another tank was supposed to be empty. The three tanks were numbered 610, 611 and 619.

It was from tank 610, which, it is estimated, contained about 43 tonnes of MIC that the leakage took place. Just across the Union Carbide plant is the shanty town of J.P. Nagar. Nearby are other localities such as Chhola Kenchi station road, Bajariya railway station, etc. It was sometime after midnight that the toxic cloud of gases descended on the unsuspecting people of Bhopal. The people first felt their eyes burning and then a choking sensation. It was one of the coldest winter nights in Bhopal. The people, who were suddenly woken up and who did not know what had hit them, started running helter-skelter. The more they ran, the more volume of gases they gulped. More dropped dead on road, wayside and bylane. Those who could not run died in bed.

Realising that something serious had happened, the Chitres grabbed a bedsheet and ran out. Outside, the scene was one of chaos. People were running helter-skelter, not knowing in which direction to escape. Some could run no more. They just fell down, vomiting, and died. For Mrs Chitre, stepping over bodies, as she ran was nightmare.

"We passed a police van with two constables who said they did not know in which direction it was safe to go. We reached a junction at the polytechnic about half a kilometre away from our home, where we decided not to go further. There, at 5a.m. a police ven arrived, telling us to go back because it was safe. No one believed them".

The Chitre said that the anxiety to flee was to great that people left their children and fled, or did not stop to pick up those who dropped off in exhaustion or because of the effects of the gas. "The scene reminded me of refugees. I thought of Hiroshima and thought that a chemical bomb had gone off", Mr Chitre said.

"At one place we saw a family give up the flight. They sat down and said, 'we will die together", Mrs Chitre said. One of their friends ran a distance of 15 km in a desperate bid to escape the gas.

From the polytechnic, the Chitres were able to telephone their friends on the other side of the town for help. When they returned to their home three days later, the pomegranate tree in their backward had turned yellow while the leaves of a peepul tree had turned black.

Mangla Ram woke up at the wailing of the siren. Normally he would have ignored it and gone to sleep again. Having lived near the factory for over ten years he is used to the siren hooting and hooting at odd hours. But on that night he got down from his bed as he felt thirsty and helped himself to a glass of water. He was still uncomfortable. Something seemed to have stuck in his throat and he was unable to breathe easily. His chest too appeared to shrinking as if pressed by a load of weight. As he opened the doors for a breath of fresh air, a grey mist wafted into the room, bringing with it an acrid and pungent smell. Mangla Ram felt nauseous. Outside a grey cloud was floating rather low over the Jaya Parkash colony, a shanty town of some 750 poor and destitute families on the north-eastern fringe of the Madhya Pradesh capital of Bhopal. When Mangla Ram came out of his house, he saw that most of the people of the colony had already woken up and many of them were running towards the factory on the other side of the highway facing the colony. Like everybody else, Mangla Ram too thought that there was a fire in the factory causing the cloud of smoke over the colony. He, along with a number of other people, looked all over the place for signs of fire but there was none. Someone noticed a thin stream of smoke emanating from a tangle of pipes and tubes.

With the cloud of acrid smoke getting denser, the atmosphere became suffocating. Breathing became an effort for everybody. Instinctively everyone ran for their homes, each rubbing his eyes vigorously to overcome the irritation caused by the smoke. It was as if someone had thrown pepper into their eyes. Mangla Ram's eyes too were moist and bloodshot. On reaching home he found his wife writhing in pain with her tongue protruding out of her foaming mouth. His two children, a son and a daughter, were also in great distress and crying. Mangla Ram splashed some water on his wife's face, wrapped her with a bedsheet, put her on his shoulder and walked out. The children ran after him.

It was a sight that Mangla Ram will never forget. An entire settlement was scampering out of their homes running south-west, towards the city centre without really knowing where to go or what to do. Many collapsed on the way, some for ever. Children vomited blood. Pregnant women stumbled and fell on the ground crying in pain and bleeding profusely. With the grey clouds of death chasing them their fears turned into panic. Relatives did not wait to pick up the bodies of those they loved and were alive only moments ago. Children got separated from their partnts, husbands from their wives and brothers from their sisters, in the mad rush to run away from the clouds. Many were trampled to death. As a terrified and sick populace moved forward, more people—the residents of neighbouring Chhola Road, Tilla Jamalpura, Sindhi Colony, Railway Colony and Chandbad settlements—joined them. The resourceful and the affluent had already fled in whatever transport they could manage to secure. Only the poor were left behind. Mangla Ram paused to catch his breath as the load he was carrying on his shoulder gained weight with every step. He waited for some time for his children to catch up with him but there was no trace of them. Apparently they were lost. As he himself was feeling sick, Mangla Ram decided to move without waiting for his children any more.

On the road, he found that some of his neighbours had stopped a truck and were boarding it. He climbed and put his sick wife on the floor. The whole floor of the truck was covered with sick people—men, women, and children—barely moving and pressed against one another. In the scant space around them were those who were still able to stand on their feet. The truck took about five minutes to reach the nearest hospital. Hamidia Hospital which incidentally is the largest in the state. Mangla did not know at what time he had left his home, some three kilometres away. But when he arrived at the hospital it was 2.30 a.m. By then the hospital had received more people than it could accommodate. All or most of them were in a critical condition gasping for breath. As Mangla Ram placed his wife on the steps at the entrance of the hospital she hardly moved. Searching for doctors he found there were very few and all were very busy. Finally, when a young doctor came and lifted his wife's hand to feel her pulse, it was already stiff and cold. The doctor covered her face with the sheet she was wrapped in and he walked away.

Down the corridor so many corpses lay one next to the other that Mangla Ram even forgot to weep.

But the actual magnitude of the tragedy unfolded as dawn broke out over Bhopal city. All hospitals and private clinics overflowed with victims pouring in a continuous stream. Tents were pitched along the roads leading to the hospitals and on the hospital lawns to accommodate the sick and the dying. They lay groaning, gasping and comitting as doctors put them on drips. The more seriously affected among them were given oxygen but soon it became evident that the doctors were fighting a losing battle as bodies started piling up in the morgues. As the spaces in the morgues were taken up, the bodies were dumped on the open grounds outside for their identification by relatives who stood in long winding queues patiently awaiting their turn. Others who hoped they would be safe in their homes died in their beds. At the railway station 25 of the railway staff including the station superintendent, Harish Dhruve, who had come for duty, were choked to death. All the beggars and urchins who spent the night on the station platforms that day, died. The city of Bhopal turned into a city of corpses—littered with bodies and bloated carcasses, dead cats, dogs and cattle. The fumes travelled as far as New Bhopal, where the Governor and the ministers live, and spread in a ten km. radius affecting a quarter of the capital's 8,00,000 inhabitants. Men, women and children trekked up to 20 km. to escape the fumes. Policemen deserted their posts. Many senior politicians and officers drove away to Sehore, 30 km. from Bhopal. The well-to-do travelled even further in cars and hired vehicles, emptying the city,s petrol pumps for the night. Bhopal was deserted.

In the factory, Shakeel Ibrahim Quraishy, the plant supervisor tried his hardest to plug the leakage. It took him an hour to do so. By then, around 30 metric tonnes of MIC had escaped, and even as Shakeel was climbing down the ladder from the tank, death was in the air. Suddenly, he slipped, fell from the ladder, broke his leg and inhaled the gas.

For about half an hour, none of this affected anybody in my house. All of us—my wife, Anita, my three-and-a-half-year-old son, Vijay, my 8-month-old daughter, Indira, and the servant girl—were fast asleep. At 1.30 a.m., Indira woke up, crying and coughing. My wife too began coughing and soon all of us were awake. I heard the siren and sensed that there had been a leakage from the Carbide plant. I knew the factory well—I used to visit it every week for printing work. My house is only 2.5 kilometres to the right of the plant, and I realised that we were in danger.

Sensing that the gas had entered my bedroom, I gathered the family in the drawing room which seemed safer, and sealed all the inlets. Even so, our throats were itching, and our eyes burned and watered. We waited for a while, and then, when the gas did not seem to be going away, I decided to ring the highest official of the factory at home. His wife took the call:

"Hello, I want to speak to Mr. M."

"He has gone to the factory."

"There is a lot of gas in our areas. We are choking. What to do?"

"You move to a safer place," she advised, "everyone is safe in the factory." Then she put down the phone. I thought it wasn't serious.

"Let us run away, otherwise the kids will die," my wife urged.

"Shut up," I said, "don't panic. I have sealed the room. Not much gas can leak in."

Later, I too got a bit scared. I went into the kitchen which had open ventilators. This room had plenty of gas. I stepped in and bolted the door.

"If we go out we will all perish," I told my wife. I rang up my sister, who lived in the BHEL township about 12 kilometres away. "Hello, I am Thomas. Poisonous gas is leaking from the Union Carbide Factory. We are affected slightly." She was sleepy. "You all come over here," she told me. "Let's see," I told her and put down the phone.

The kids were not coughing now, but our eyes were troubling us badly. Another phone call came from my sister enquiring about our safety. We waited, tense.

Jaiprakash Nagar 1.30 a.m. This colony is about 600 metres to the left of the factory. It has about 700 huts with a population of over 5,000 people. Most of the man are labourers, cart-pullers, kabadi wallas (Junk sellers and buyers), and some dairy owners. A poor, but fun-loving people, they were fast asleep after a hard day's work and the evening's entertainment of music.

Kallu (45) woke up coughing. He knew it was gas from the factory. He came out and found a lot of people running. He also ran, leaving behind his old mother and 13 milch buffaloes. He reached the Shamla hills, where most of the state ministers live. The buffaloes were restless. The helpless animals pulled at their chains, which tied them to the hooks firmly embedded in the ground. They fell one by one and choked to death. His mother put a blanket over her face and lay still; perhaps she was unconscious. She survived, miraculously.

Sarun Mian also owned a couple of buffaloes. Along with his wife, and a three-day-old baby, he ran out, but it was too late. His wife fell down on the road along with the baby. . . Sarun kept running, as there were hundreds of people behind, pushing him. His wife and baby died. Later, the government gave a number to the dead body of his wife: it was 505. The dead were, identified only by numbers that day.

In the same colony lived Prem. He had helped many of his neighbours with jobs and money, and he stayed in a pukka house with his wife and a six-year-old son. They too woke up coughing. The couple opened the door and heard others shouting, "Run. .run." They ran towards the nearby Bhopal railway station. The atmosphere was thick, white and filled with smoke. Only when they reached the station did they realise that their only son was not with them. They rushed back towards their house. He wasn't at home. As they searched, his wife fell down. She had inhaled too much of the toxic MIC. Prem tried to lift her, but he was too exhausted. Neither dead bodies were found outside their house the next day. Ironically, their son survived and was found at the railway station.

Ayub Khan and his family woke up that night, their eyes moist and red. They were coughing. They too, started running.

They didn't know where, nobody cared about anybody else. Today his sons Anwar (20) and Ansar (16) are missing. "Are they alive?" he asked.

"We were running and running; many fell down vomiting and were trampled by the others."

Cart-puller Nazir had nine children. Six of them: Sagir, Mateen, Muin, Shanaz, and Saheen all aged between 10 and 24, are missing. "Are they dead?" he asked, looking towards the vultures hovering above the carcasses of buffaloes, that were still to be cleared. Gafoor Khan, a labourer, said, "My wife and son are missing. I have helped bury 25-year-old Nazir, 20-year-old Sagira and 12-year-old Saleem. They were all my relatives."

Hanifa ran off to the hill where the central jail is situated. "The road was full of people, trucks, tempos, cycles and autorickshaws," she said. She returned home on the fourth and found her brother and his family of five lying dead in their house. They had died in their sleep.

By December 4, only about 300 people had returned to this colony that originally housed 5,000 inhabitants. The rest were dead or missing. A hushed silence huge over the colony when I visited it on the fourth. Several dead bodies were still to be cleared. Dead cattle lay all over the place.

At about the same time, Mr. and Mrs Chacko woke up. He was the Personal Assistant to a top official at Straw Products, on Cholla Road. He was one of the very few lucky ones—he had company quarters in the factory premises. Unfortunately these quarters were just 500 metres away from Union Carbide. Both of them inhaled the gas and woke up coughing, with red and watery eyes. Their only son was sleeping. "We must wake up the other officials of the factory," he told his wife. Both of them went from door to door, knocking and urging the people to run away. Many of the officials were unaware of the gas, as their bedrooms were well sealed, due to the extreme cold outside. They were fast asleep. The Chackos had to knock repeatedly on many doors, losing valuable time. The killer gas took its toll. Chacko fell down unconscious and died. His wife is in hospital in a serious condition. Their son survived.

Just behind this factory lived Jacob. He came out in a daze, having already inhaled a lot of the gas, took his scooter and sped off to JP Hospital, about seven kilometres away, inhaling more gas all the way. He was admitted to hospital but died that evening. His dead body was brought to the Hamidia Hospital mortuary. A number was attached to the body. There were many others from Straw Products colony who perished that day.

Sindhi Colony, about a kilometre ahead of Union Carbide, is a colony of prosperous Sindhis. At about two a.m., Dayaram and his family awakened. They smelt the gas. Instead of locking himself up inside, he ran out yelling. His wife followed. There was an unending stream of people on the road. There were also trucks, tempos, cars and jeeps full of people, fleeing. As they ran, a jeep knocked his wife down. He lost his head, running after the jeep, howling abuses. Later, he realised his folly and came back looking for his wife. What he saw was nauseating. His wife's body was mutilated beyond recognition. Hundreds of people and some vehicles had swept over her. He started running, shouting. He reached Hamidia Hospital, the biggest hospital in Bhopal, but the least prepared for such an emergency. There were barely half a dozen junior doctors on duty along with a dozen nurses. Medicines and water were scarce. Oxygen cylinders were in short supply. Some of the hospital staff were also affected. Yet, thousands of people poured in any many of them died. Dayaram was one of them. He was numbered 125 and his body was packed away in the mortuary between the other dead-bodies, like sardines in a tin.

The Bhopal railway station Superintendent, S S Dhurve was wide awake and and on the telephone when the gas entered his room and left him dead immediately. The Bhopal railway station, about 1.5 kilometres towards the left of the factory, was hit by the gas as soon as it started leaking. Hundreds of beggars, porters, passengers, and railway staff fell down unconscious and died. They never knew what happened.

In Chandbarh colony, towards the left of the railway station, Mangat Ram, a young man, slept. His wife got up coughing and woke him up. It was 1.30 a.m. "Some thing is wrong," she said. Their two sons, aged four and two, were sleeping. As he opened the door of his hut, white smoke came billowing in. His wife collapsed. He carried her out. Everyone was running. They could see nothing. They boarded a truck, going towards Obaidullahgunj. He also collapsed. His wife died later. When he returned, his children were missing. Perhaps they were among the 200 unidentified children cremated or buried at the Cholla cremation ground.

Cholla Kenchi is a colony of over 300 huts and 3,000 people, just behind the Straw Products factory and close to Union Carbide. Anand Kumar (40), his wife (35), five children and four close relatives were fast asleep in their hut. The greyish, poisonous MIC rushed in through the wide opening beneath their asbestos roof. None of them ever woke up. They died in their sleep. M.K. Singh, his wife and three-year-old son ran away to Niashat Pura. Their eyes were badly damaged. They returned home by the afternoon. Little Atul was hungry. His parents offered him bread which had been exposed to the deadly gas. He died three hours later.

This was the general pattern of events on the fateful black morning of December 3, 1984. Thousands and thousands of people fled to distant places like Indore, Dewas, Vidisha, Itarsi, Hoshangabad and Jabalpur. Many perished there. Thousands have been lost.

My home, five a.m. We had sat up all night in our drawing room, too tense to move. I had decided against accepting my sister's offer of shelter—the air outside was probably unsafe. But now, the gas had probably dispersed. Tentatively, I went out and inhaled. It seemed as though the obnoxious smell of poison gas had gone.

We decided to venture out on my motorbike. Thousands of people were still running away. We reached the higher and safer Hamidia Hospital area where survivors and the injured offered solace to each other. After an hour, seeing the death and horror, I decided that we were better off at home", a correspondent said.

That day, death had attacked the city of Bhopal. Those who were well off and able, ran away in vehicles or bolted themselves into their houses. They survived. Those who were poor and weak could not run away and they did not have a safe house to stay back in. They perished. The main hospital was littered with dead bodies and the seriously ill. Doctors and medicines proved grossly inadequate.

At the Cholla Road crematorium, Dayaram Manwani, a RSS volunteer, spoke: "We are eight volunteers serving here. we have cremated over 600, including about 200 children. The government has provided the wood and kerosene for this. The Muslims are being buried at the three burial grounds in the heart of the city."

Over 4000 people have died in the disaster, about 25,000 are hospitalised and an equal number are missing. About 4,000 cattle have been lost. It has been the worst disaster in the chemical history of the world, second only to the atomic explosions in Hiroshima and Nagasaki. Experts are of the opinion that if the safety valve had not worked then the underground storage tank would have burst, wiping out all traces of life in a radius of 100 kilometres. All shudder at the thought. As one would have lived to tell this tale.

—(*Imprint*, January 1985).

Mr. G. looks completely shrivelled up, his old sprightliness is gone. Is this the Mr. G., who has never lost his quiet self-confidence, his moral strength, his dignity in the last 64 years. Is he cracking up too. He didn't smile even once, but was polite, if monosyllabic throughout—except when I asked after his son who lives in Delhi . . . . "I want to quit Bhopal. I don't feel safe here, I never will. They have ruined Bhopal, the city of lakes and hills, the quiet place where I'd thought I'd retire, my home-town. They have finished it for good."

Mr. G. woke up coughing violently shortly after one a.m., on December 3, cursing himself for his weakness for the five to eight cigarettes he still allows himself. The phone rang before he had switched on the night lamp, but just after his wife too woke up complaining of severe eye irritation. It was his friend from the secretariat, telling him that poisonous gas has leaked from the Union Carbide plant and that he should run, as fast as he could with his rheumatic wife their daughter-in-law and two-year-old grandchild.

By the time Mr. G, had finished talking, everyone in this middle class colony, about six km from the plant, was up and rushing out of their homes. Mr. G.'s family panicked as their eyes burned and their throats choked. They fled, Mr. G. without his high-powered (minus eight) glasses and barefoot, the others still in their bed-clothes, all packed in a neighbour's car. They left behind the neighbour's two sons, one of whom had fainted upon breathing the gas, while opening the garage door. . . The younger of the two men managed to take his brother to a hospital in new Bhopal. . . He has become unconscious once again, and is now half dehydrated thanks to Lasix.

Mr. G. has not yet come out of the trauma. Will he ever? The sheer indignity of what he had to go through, his granddaughter's worsening cough (the private doctor he managed to find after two days of search suspects pneumonia), and the insecurity he feels are incompatible with the image this honest, upright, middle class man has of himself and his world. . .

Ran into Ramlal at the Kali Parade relief camp across the Carbide factory. Ramlal works as a day labourer, and lives just behind the plant in a bustee. He looks very ill, his breathing is laboured. The doctor gives him Tetracycline and sends him home. I wonder if he will survive. Ramlal is angry. "The saabs of Union Carbide never sounded the siren," he says. "If they had done so before or just after the gas leaked out, we would have known what to do. We'd at least have had a chance to run. We have had phosgene leaks and alarms in the past. I lost my wife and two sons only because they didn't warn us. . . These people are criminals, butchers, worse than murderers. . ."

Like natural disasters, man-made ones also seem to have a preference for the poor. Over 90 per cent of the worst affected people are the poor living in the close vicinity of Bhopal's industrial area. They ran as fast as they could, but how could they overtake the cars and scooters of the rich and the upper middle class? And what is the ownership of a telephone if not the best chance that you will be warned by friends, relatives, neighbours of impending disaster? By four or five in the morning when the police went into Chhola, Jaiprakash Nagar and Tila Jamalpura, all working class localities, their job was clearly defined; Open doors to one-room tenants and pull out bodies—five, six seven.

After much waiting and several telephone calls, I managed to get into the Union Carbide guest house. I told the two top men of Carbide that I was tired of being told that they themselves did not know what had happened, because, you see, the CBI had seized all plant records. Didn't they know from their own employees, who were in constant touch with them? No, we wish we could tell you, we're not really sure what leaked, and how much . . . Didn't they know that the seized records showed very little—the log-book was basically blank as far as the third shift (during which the accident occurred) went? The hotel staff want us to eat early because they are four short—missing relatives, one death . . . the nightmare sinks into one's consciousness, one struggles to keep the images out, to anaesthetise oneself.

On December 3, Swaraj Puri, Bhopal's S.P. was woken up by a town inspector's call telling him that people in Chhola, about two km away from the plant, were fleeing and there was some gas leak . . . Puri rushed to the police control room, but not before. When he arrived there at 1.25 a.m. there was chaos. The duty staff were coughing violently and rubbing their eyes, as the acid smell of gas got stronger. Puri put two of his men on the job of calling the Union Carbide plant to find out what had happened, which gas had leaked, what preventive or curative measures were to be taken.

Between 1.25 and 2.10 a.m., he got through three times to the plant. Twice he was told "Everything is OK," once "We don't know what has happened, Sir" before the phone was banged down at the other end. Meanwhile, reports poured in of over 1,50,000 people having fled their homes in futile search for safety. Puri's first response to anyone who asked him what to do was "Don't panic, stay indoors, panic will make the situation worse"—a stock reply which wouldn't have been given had he known what gas had leaked and how lethal it was.

But Union Carbide officers wouldn't call the police and no one knew what to do. Government doctors were at a loss to prescribe remedies. The rest room next to the control room provided the only form of relief possible to the lacrimating policemen. They turned on the two fans in it to full speed, gathered west towels and periodically made a quick sortie into the rest room covering their faces with the towels—back to work, dialling the plant, and sending out relief teams again.

The jumbled cameos of horror in Bhopal made up a monstrous catalogue of sufferings. Nearly half of those who died were children. School teachers called the morning roll and many were unanswered. In the poor districts near the Union Carbide plant people stood in groups, coughing and wheezing from the gas, numbed by grief, as they talked:

"We were a family of six, four died."

"The couple next door died. The wife had just given birth, to they could not run."

"I am a night watchman. I came back and my wife and two children were lying dead."

In all these cases there was only one common element—grief and death. Everyone talked of the gas having the pungent smell of chillies thrown into a hot pan. It was the acrid smell that awoke them. They thought at first that someone was cooking.

Ram Sevak Piasi's search ended. As soon as he heard of the gas death cloud at Bhopal he caught a train and journied 150 miles to find his brother and his family lived near the Union Carbide plant. He searched and questioned for days. Finally, he found his sister-in-law alive in hospital. At a police station he looked into the large poster hanging on the wall with the headline—"gas accident, unknown dead." His brother and the nephew were among them on the list. In the worst affected areas, where the white cloud of MIC was heaviest and lingered longest, there was a pervading sound of coughings. People queued at make shift, open-air-clinics and waited for their turn to be examined by the doctors.

At one of these clinics, Dr. R.K. Jain, a large and kind-hearted man in a dark jacket and tie, dispensed pills from a heap of medicine on a table and gas oxygen from a cylinder. In the past few days, he said 25 people died at his feet only, their tormented lungs giving out after the last desperate journey for help. The doctor remarked:

> "We are seeing 1,200 victims a day." At first, there was respiratory trouble, then fever, afterwards, they were coughing blood. We did not know what symptoms we would see next."

In Hamidia Hospital, the largest of seven in the city, people lie convulsed, coughing and vomiting. In the children's wards infants lie on long tables, oxygen tubes in their nostrils, struggling for life, their weary, daze-faced mothers dabbing the babies' mouth with water. The mothers stayed night and day, snatching sleep when they could. A new-born baby lies with his head in a plastic-hood. Outside the hospital were rows of green tents. emerging overflow of wards erected when the hospital was swamped with thousands of gas victims. A woman with streaming eyes said:

> "I can see nothing. I remember the smell of chillies and ran. Every one ran. We were six in the family but I do not know if others are alive or dead."

The exodus on that fateful night of terror and panic was huge. The streets were left littered with sandals. Doctors thought that some of those who fled died more quickly because the effect of running made them inhale more quickly and more and more of poisonous gas. Dr. Nagoo, the health director, met a human unending tide as he drove into town. People climbed into his car, onto the roof and bonnet and boot, as he drove them to hospital.

Brigadier. M.L. Garg, a former army officer, now works as the manager of a paper plant employing 1,200 people said:

"The people of Bhopal were frightened and angry. They were frustrated and wanted justice. This was a devastated community. People lost confidence in the authorities. Our own workers were reluctant to return. They did not believe the assurance of the authorities that the gas had gone even after a week of the gas leak."

The Brigadier said that he had been left with a tight, asthamatic feeling in his chest. His wife had been temporarily blinded:

"The gas was in the house like a yellow haze. I called on the local army unit to evacuate our night-shift workers and others. Some fled. In all, we lost 14 people.

"There had been deaths before at Union Carbide plant and this made us anxious. The plant should not have been so close to the city. Because of the leaks we had doubts about their preventive maintenance", the Brigadiar added.

A pregnant woman who had delivered a still-born child on December 8, 1984, died at the Sultania Maternity Hospital. A grim fall-out of the tragedy was that a large number of families had been entirely wiped out and scores of children rendered orphans. It was the future of these kids that looked pathetic. Bhopal disaster was India's Hiroshima.

Bhopal mortuary was littered with bodies of men, women and children. Thousands of people were ferried to hospitals in lorries, buses and cars and tents had been pitched in hospital grounds. There were hundreds of dead dogs, cats, cows, buffaloes, birds lying in the streets.

He said greater attention would also have to be paid to avoid after-effects of the poisonous gas on certain items like vegetables to make them safe for human consumption. He added that there was no risk in drinking water. Expressing his profound grief over the tragedy, particularly about the death of the little children, the Prime Minister said the worst danger was over and added that in future the government shall have to do contingency planning to meet such incidents, particularly to ensure that hospitals were well equipped to meet with situation.

Meanwhile, the Election Commission had no alternative but to postpone the 8th General Election to the Lok Sabha seat of Bhopal, especially after nearly one fourth of the city's population had left the city in panic after the disasterous leakage. Even though "Operation Faith" to get rid of the remaining MIC gas in Union Carbide plant proceeded without a hitch and people started coming back to the city, it took quite some more time before the situation returned to normal and the people could take interest in politics and elections.

A gap of nearly one month from the original date of election also enabled political parties, the candidates and the electorates to think dispassionately about the calamity. The gas leakage deaths were no doubt the most important among issues in the general election campaigns when it resumed.

The Congress(I) blamed the Janata Party government during whose regime the plant started production and the opposition parties pointed out that despite a number of accidents of a serious nature of its operations, the Congress (I) government in the Centre and the states not only permitted the plant to function unhindered but also allowed the residential area to come up within its vicinity. Of course, the Congress (I) candidate won the election from the Bhopal constituency.

What happened in Bhopal was a warning to India on the road to industrialisation and Western world already industrialised like the US, the Soviet Union, and the rest of the developing nations following the course of industrialisation. The lethal danger was not from the atom bomb, though that was real enough. But the one who died, or were marked with gas burns on their bodies, were not the victims of a bomb-attack. They were the victims of an invisible gas that was leaking through a defective value.

Union Carbide officials in charge, frankly, knew nothing about the chemistry of the poison gas that was leaking out of the plant. Union Carbide, a multinational undertaking was located only a kilometre off the main city. The distance was too small from where they were making virtually a poison gas.

One leakage was enough to spread poisonous but invisible vapour over a wide area. It was poisonous to reach several kilometres and lethal enough to kill or at least incapacitate lakhs of people. Indeed MIC proved deadlier. The effects of Methyl Isocyanate in Bhopal were more deleterious and extensive than doctors and toxicologists had originally thought of, both as an immediate killer and as a slow and long acting poison which affected a number of organs in the human body and damaged them, in some cases beyond repair.

The most devastating evidence of the frightful health effects of MIC to which about 2,50,000 people were affected in Bhopal when 40,000 kilos of the toxin was released from Union Carbide pesticides plant on December 3, 1984, was provided by a number of recent medical observations.

These included chest pain, breathlessness and cough (on account of extensive lung and respiratory tract damage), difficulty in concentrating, depression, anxiety, changes of consciousness, instability and poor memory due to hearological damages as well as abdominal pain, vomiting and nausea due to adverse effects on the gastro-intestinal tract.

The victims showed respiratory symptoms such as dyspnoea, tightness in the chest, irritation in the throat, severe cough and a degradation of the lung function. Their blood showed the presence of high quantities of methaemoglobin and Carboxyheamoglobin. On the basis of the known effects on the human body of inhaling some types of isocyanides, babies to be born in the next months of gas victims in Bhopal would face brain damages. This would be the long term genetic effects of the inhalation of methyl isocyanate (MIC).

Even the foetuses and newly born babies fell prey to the deadly MIC gas which leaked from the factory. One child was born with holes where his eyes should have been. Many who survived were diseased or dying. The other effects of syndromes caused by exposure to the deadly toxic gas that were cause for further alarm and shock. Soon after December 3, 1984 the effect of the gas on about 1000 pregnant women living in the 16 areas of Bhopal which had been declared as "gas affected" was feared.

The negative effects of MIC gas on pregnant women could be most apparent in children born around May-June, 1985, since formation of the child was nearly complete after the first trimester (three months). But the records of births and deaths of infants in Bhopal proved our expectations.

A team of scientists who had conducted studies in Bhopal in the wake of the gas tragedy presented its initial findings at the Science Congress session at Lucknow on January 6, 1985. The scientists included Dr. (Mrs) S.V. Chandra and Dr. M.U. Beg of the Industrial Toxicology Research Centre at Lucknow. However, much of what emerged from the K.E.M. hospital in Bombay study fell into a pattern noted earlier in past cases of isocyanate poisoning followed in Britain. Two of these studies related to a leak of some 4000 litre (or about one tenth the quantity released in Bhopal) of tolerance di-isocyanate from tanks in a British polymethane plant. Thirty-five firemen were exposed to the TDI. They were closely monitored for over four years.

The majority showed both neurological and respiratory complications. The symptoms set in slowly and, in many cases after a considerable delay. As many as 43 per cent, of the men showed signs of long-term damage and a decline in lung function as long as four years after exposure. In many cases, there was an aggravation of bronchitis and other respiratory diseases. Neurological symptoms tended to persist for four years. These included ataxia of gait, parasthemia, impotence, confusion, difficulty in concentrating, poor memory and depression.

There was reason to believe that on account of being more volatile and reactive; MIC could be more dangerous than TDI. Yet TDI was itself a very potent poison, even a concentration of only fine parts per billion in the working environment (for only 15 per cent of the time in an eight hour weighted exposure) could lead to damages in lung function. It was not only the lungs and the central nervous systems that had shown signs of long term damages. Many of the organs, including the kidneys, liver, spleen and even the stomach, had been noted to be involved. Plainly, the challenge posed by MIC to the Indian medical community could be as formidable as the substance was deadly. Apparently, the eyes of the gas-victims were burning and watering, but hurt more. Their vision blurred, they were all short-sighted. Several of the affected children had bruises around their lips—from constant chewing because of the suffocation.

Adults and children alike were breathing not with their chests but with their abdomens, and it was laboured—in sharp gasps. Their pulse was low. Several complained of paralysis of the lower limbs, and generally a weakness of the extremists. They were scantly clothed. They felt such intense heat that they had cast off their clothes. So many of their corpses were brought in without a scrap of dress on them. One odd feature that immediately caught the eyes was the number of corpses that came in stark naked. The weather on those nights were not warm for Bhopal.

It was for Bhopal a normal winter temperature. Between 8.30 and 9.00 a.m. the bodies began to be brought into the morgue in hundreds. Outside the hospital, the victims poured into campus in waves. The out-patients' department recorded a figure of well over 12,000 on that fateful day of December 3, 1984. Across the road was Jayaprakash Nagar, a huge slum colony, that ran almost the entire length of Union Carbide's boundary wall in Bhopal. Only corpses could be found there. It was obvious that the immediate task was the urgent removal of the dead bodies and the carcasses of animals to avert an epidemic. With the help of the Army and police the corpses were cleared. It was aweful. Knock any door and all you could find were dead bodies.

Behind the human dimensions of Bhopal tragedy, the whole region became the latest ecological wasteland. As some one said, just like Seveso in Italy and Minamonta in Japan, where the grim consequences of industrialisation were savagely wrecked upon an unsuspecting population. At a cost which was incalculable in terms of human suffering and anguish. At a price which, chillingly continued to be exacted from the progeny of the survivors.

Medical experts in Bhopal were waging a losing battle against an unknown disesase and an unidentified poison. The first case of deformed suspect because of MIC exposure was of a child born in Bilaspur's Sardar Vallabh Bhai Patel Hospital. The mother of the child had been near Union Carbide factory at the time of leakage and had been exposed to MIC. The child born to her by a caesarean operation weighed 4.5 kg and had two gaping holes instead of eyes. The baby had no fingers and toes and skin looked acorched. Between 20 and 30 infants were born every day in the Sultania Janana Hospital which was among the largest women's hospitals with deformity of one kind or the other. According to gynaecologists and midwives at the hospital's maternity ward, the cases of premature or retarded birth among children being born were abnormally high in numbers. As many as 50 per cent of the children being born were either underweight, premature or had some physical deformity. Many of the children born did not survive. And where the baby survived it was in a precarious condition and had to be sent for emergency treatment to either the Hamidia Hospital or the Medical College Hospital.

The Sunday cover story reports:[1]

Sushila, aged 20, gave birth to a still-born child on March 29, 1985. A resident of Shahjahanabad colony, Sushila had a different pregnancy ever since she was exposed to the MIC gas. She complained of restlessness, fatigue and breathlessness. After a prolonged labour a frail baby girl was born to Sushila. Before the baby's birth, senior gynaecologists had examined the mother with ultra-sound equipment. The child that was born had a deformed head formation. The back of the child's head was an extended cyst-like formation and part of the skull was missing. This child too had scars and burn marks all over its body. The baby died the day it was born, said Sushila:

> "It is the gas that had killed my child. Ever since the time I inhaled the gas, the progress of the child was not normal. But the doctors told me nothing. They simply gave me medicines and injections."

1. *Sunday*, 7-13 April 1985.

The Sultania hospital rccord stated the child was born after 36 weeks of pregnancy after an abnormally long labour. The mother of the child was "MIC-exposed." A look at the register revealed many more abnormal births to MIC exposed women in the past three months. There were confirmed reports of births of other deformed children at Sultania Hospital. Of the 3,000 gas-affected Bhopal women examined by Specialists, 600 were found to have delivered still-born babies. Of the five hundred babies, four hundred died within few days. Among the survivors, a number of them suffer from deformities. These apart 60 per cent of all pregnant women suffered abortions. This has once again surcharged the atmosphere in Bhopal, in which background the counsel of UCIL has pleaded for a change of venue of the probe into the gas tragedy by Justice N.K. Singh.

On February 20, 1985, for example, three unnatural births took place in the maternity wards. One child was born with a skull, another with unformed limbs; while the third one was born two months before completion of pregnancy. The last child had patches of blue skin and was born without hands. Reports of still born or abnormal births had also been received from private nursing homes and maternity clinics in Bhopal. Nurses at the Ghoois Nursery Home and Termination of Pregnancy Centre, Kalbala Road hesitantly disclosed details of such births. There was a high percentage of premature births and in the past two weeks at least three children died because of deformity. The hospital gynaecologists had been advising the pregnant mothers exposed to MIC effect to have their pregnancy terminated. But those pregnant mothers who did not visit hospitals became the victims of MIC effects. A report published by the Indian Council of Medical Research (ICMR) on the "Health effects of toxic gas exposures at Bhopal in January 1985 had clearly stated a high percentage of unnatural births because of MIC exposures. The report indicated:

> "During the acute period 36 pregnant women were hospitalised for symptoms of respiratory distress. Five of them aborted, of whom three were definitely attributable to MIC effect. Of the 30 antenatal cases, ten were seriously affected and four of them had normal deliveries."

It was suggested in the ICMR report that a five-year follow-up s tudy be undertaken to find out morbidity referable to CVS and respiratory systems. There is a need to identify abortion rate among MIC-exposed women, the intrauterine growth retardation among the next generation that the ICMR scientists suspect long term effects of MIC gas on pregnant women and infants.

Clinics and dispensaries situated close to the Union Carbide plant, along side Jayaprakash Nagar, Chhola Kenchi and Railway colony at Bhopal, which were the worst hit localities, had an increased number of women patients. Thousands of women exposed to MIC and undergoing treatment for exposures had complained of varieties in menstrual cycles. Some women complained of menstruation five times in the course of one month. Lactating mothers reportedly informed doctors about a reduction in milk and a change in the taste of the mil.k On the other hand, MIC affected men had together with general fatigue and poor body health complained of impotency.

Many victims were treated at the New 30-bed hospital set up to meet the unprecedented rise in cases of MIC exposure. Though the hospital (once the residence of the police DIG Bhopal) was equipped with respirators and provided facilities in skinography and pathelogical tests, prevailing conditions at the hospital were appalling. Over 500 cases of MIC effect were attended to by the small team of doctors at the hospital. In the last week of March 1985, the 30-bed hospital recorded a rise in the number of patients treated: more cases of MIC afflictions were reported. On March 29, 1985 the hospital treated as high as 900 cases of MIC-exposed victims.

The medical specialist treating MIC cases at the hospital, Dr. (Mrs) S.P. Verma said that treatment was given to victims for respiratory disorder, fatigue, loss of appetite and pulmonary complications. Serious cases were given regular doses of sodium thiosulphate. She also complained that a large number of the gas-affected victims coming from areas close to the Union Carbide plant were suffering from serious mental disorders. An old porter who had nearly been blinded by the gas and was a witness to a previous gas leakage and a major fiee at the Union

Carbide plant close to his house was in a state of mental trauma. Incapacitated and sick, he saw the Union Carbide plant as a monster that wrecked his life and many lakhs of others.

Of the two lakhs Bhopal gas leakage victims underwent treatment, 10 to 12 per cent had psychiatric complaints or suffered from acute mental depression. This meant 20,000 victims were on the point of mental derangement since such treatment facilities were not made available. A team of psychiatrics had visited Bhopal in December 1984 immediately after the gas leakage tragedy and had made important observations about the mental health of the victims. The team included Prof. D.B. Sethi, chief psychiatrist at the K.G. Medical College, Lucknow and Dr. D. Shrinivasa Moor of the National Institute of Mental Health and Neurosciences, Bangalore. The report of the team had concluded with these words:

> "Symptoms of anxiety and depression are foremost among the patients but it is difficult to say if these are just emotional reactions on the effect of the character."

Instances of mental disability among gas victims rose. In the orphanage set up by the government, children had received early psychiatric treatment. But the children were in a state of deep shock. They had been separated from the family and now they were on the verge of lunacy. The young boys had been impaired of their speech since the day the gas escaped. For thousands, who lost their near and dear ones due to the leakage of gas were in the grip of nightmare. The tragedy of Bhopal's victims was that having fallen prey to MIC, they became the target of medical experiments and haphazards treatment. They had become guinea-pigs.

Soon after December 2/3, 1984, the main point of contention was whether the victims should be administered sodium thiosulphate, an antidote for cyanide poisoning, or not. This controversy had not yet been sorted out till date even though sodium thiosulphate was administered to several victims. The remote possibility of there being cyanide poisoning in MIC-affected persons was discovered by Dr. H. Chandra, Head of the Forensic Department, Hamidia Hospital where autopsies of all persons killed by MIC exposure were conducted. The pronosis of Dr. Heeresh Chandra was that the blood of those persons killed was cherry-red, as in the case of cyanide poisoning. In 1976 report on methyl isocynate, printed by the Union Carbide, it was stated as thermal decomposition of MIC produced by hydrogen cyanide.

The "miraculous antidote" was eventually administered to critical MIC cases on February 14, 1985; more than two months after thousands had died an instant death (80 per cent of the deaths in Bhopal following the gas leak took place outside hospitals, only 438 deaths were recorded in various hospitals on December 3 and 4, 1984). Hundreds of lives could have been saved if the injections had been administered to victims at an earlier stage. But what was more surprising was that even after the reaction of sodium thiosulphate on at least 50 per cent of the victims had been found to be satisfactory, the intravenous injection was not administercd in a regulated manner.

Instead of the stipulated doses of 12.5 gms, some victims were given doses of only 1.5 gms of the antidote. At least 1000 persons who were among the worst affected should have been given sodium thiosulphate regularly. However, not more than 500 victims were reported to have received it. With limited stocks available (20,000 vials were received in January, 1985) still for many victims, even since the gas escaped it had become a living death for them.

Doctors at Hamidia hospital who made inquiries with the chief medical officer of Union Carbide, were told that MIC was an irritant, but not fatal. Dr. Chandra had has suspicions. There was not much on the gas in his toxicological manuals, but he learnt that it was highly toxic and suspected from its chemical composition, that in the human body it would break down into cyanide and other toxic amines. If it was cyanide, then of course, the only antidote that could be administered was sodium thiosulphate.

The Union Carbide management had said it was phosgene, not MIC and as such it was toxic. The Works Manager of Union Carbide who met a *Navabharat Times* correspondent on the site at 4.30 a.m. on December 3, 1984 assured him that MIC was not poisonous. The issue was not whether something was true or false. It was a question not of science but of lives. It was, however, clear that the deadly gas that leaked out of Union Carbide plant was phosgene mixed MIC. It was toxic. The crucial upshot of phosgene theory was that it reinforced symptomatic treatment. Each symptom was dealt with separately:

eye drops for the eyes

—antibiotics to prevent infections,

—antacids for the stomach.

There was no attempt to purge the blood of the toxin, which continued to ravage the organism from within. The phosgene theory, however, as some doctors and other investigators pointed out, had larger implications. It successfully diverted attention from the toxicity of MIC, a base chemical on which Union Carbide as a company built its chain of pesticides products.

This was apparent in the bright red, almost pink colour of the blood, and the unpleasant odour given off when the lungs were cut open—the oil of bitter almonds, the smell of cyanide. Also indicated was concentration, thickening of the blood. The laboured abdominal breathing of the survivors seemed also to be a pointer. In destroying two crucial enzymes in the cells, cytochrome coxidase, and rhodinase, cyanide seriously inhibits cellular respiration—that is, the ability of the blood to absorb oxygen. This could stimulate respiration, because the chemoreceptive cells respond as they do to decreased oxygen.

The cyanide findings were disputed by Union Carbide officials on the grounds that cyanide was not used to make MIC. Dr. Chandra, however, pointed to the presence of cyanide radical in its composition. The isocyanate would react with water in the body to break up into dydrocyanide. Dr. Chandra informed the other department heads who were in control of the hospital wards, of his findings, and of the recommended antidote—sodium thiosulphate. This was rejected.

It became increasingly evident that this was not phosgene, on which the Forensic and toxicology department, Hamidia hospital had done some research. It was the department that had done the postmortem on Union Carbide workers who had died of exposure to the gas. Phosgene caused continuous coughing; but the abdominal gasping was altogether different. People exposed to phosgene had only to be moved out of a polluted environment into fresh air before they showed signs of improving. In cases of severe exposure, only symptomatic treatment was recommended, since it did not enter the bloodstream, MIC did. By that afternoon, the first autopsis had been performed. In many cases, the blood was purpled, the lungs were filled with their own secretions. Several of these cases could possibly have died of "drowning". The tracheas wore so dry that the mucus flanked off at a touch:

> "Sometimes the blood was so thick that if you dipped your finger in it, and lifted it, it would come off like a wire".

Immediate administration of sodium thiosulphate to all victims was recommended if cyanide poisoning had taken place. The chairman of Union Carbide Corporation (UCC), Warren Anderson, who visited Bhopal hours after the gas escaped brought with him sufficient supply of the drug. A German toxicologist, Dr. Max Daunderer arrived in Bhopal on December 7, 1984 with large supplies of sodium thiosulphate. Still the antidote was not given to the victims. A few days later, when the news of suspected cyanide poisoning in Bhopal was known internationally, a representative of UCC himself recommended the use of sodium thiosulphate. Within a week of the tragedy, Dr. Bipin Avashia, an Indian working as medical director of the Charlson plant of Union Carbide, USA, sent a telex message to the Bombay, Delhi and Bhopal offices of the company:

> "If cyanide poisoning is suspected use amply nitrate. If no effect use sodium nitrate 0.3 gms and sodium thiosulphate 12.5 gms", says Dr. Chandra. The telex message was clear

admission that the Union Carbide's parent company itself suspected MIC to have decomposed into cyanide, the reason for the cherry-red blood in the dead persons' bodies. The Bhopal plant was itself to launch from December 1984 a range of new products, including a formulation identical to Baygon, the household pests spray, all of which were based on MIC. Many chemical corporations are raised on the exploitation of a single base chemical, and the toxic consequences of MIC threaten Union Carbide Corporations around the globe.

Though methyl isocyanate was the villain of the tragedy, initial reports, even from expert toxicologists, were often in conflict as some suspected the killer gas to be phosgene. This gas is found from methyl isocyanate and shot to notoriety when it was used by the German Gas Corporation across the British trenche on December 19, 1915. What contributed to this confusion was the fact that very little was known about methyl isocyanate poisoning even by the experts. The Phosgene ($COCl_2$) was responsible for 80 per cent of the gas casualties then. Ten times more toxic than chlorine, it gives no warning to the victims. Smelling as it does of fresh hay, the effect shows up after an hour of inhalation.

Also supportive was the reasoning that with a boiling point of 32°C the liquid methyl isocyanate (or MIC, as it came to be referred) could not have vapourised on a night when the ambient temperature was as low as 14°C. Phosgene, on the other hand, turns into gas at a temperature of barely 8°C. This meant reasoning, however, in favour of Phosgene fell to pieces against the background that the MIC stored in stainless steel tanks had inexplicably heated beyond 40°C increasing pressure tremendously within the tank, triggering the disaster.

Until the Phosgene theory gave way to MIC, experts debated the question of an antidote seriously for the Bhopal gas tragedy but people died like flies. What was amazing was that such disinformation could gain ground in the face of the death toll that doubled and crossed the 3000 mark in just four days. The threshold limit value for MIC in the air, that was the limit beyond which its concentration in the atmosphere became dangerous, was 0.2 parts per million (PPM), For Phosgene it was much less 0.1 ppm.

Below 39°C MIC is in liquid form, stored under light pressure and a nitrogen blanket. However, in the night of December 3, 1984, it was a gas, not as a liquid, that 40 tonnes of it escaped into the atmosphere. Because of some inadequate safety systems, the temperature in the tank was much beyond 39°C which raised pressures to the point that 40 tonnes of it was forced out into the atmosphere within the space of an hour and a half. Its behaviour at such high pressures and temperatures in the atmosphere and in human bodies was crucial.

"The occupational Health Guide-lines for Chemical Hazards", published by the US Department of Health and Human Science, documents the effects on four human beings exposed to varying concentrations of the gas from 1 to 5 minutes. At 0.4 ppm there was no effect. At 2 ppm, there was watering of the eyes, irritation of the nose and throat, which at 21 ppm became unbearable. In other words, before Bhopal, the highest known human exposure to the gas was 21 ppm for 5 minutes at the most. Given the exposure in Bhopal the relevant ratio was probably not MIC to air but air to MIC. It was the tests on animals, also reported in the same volume, that were closer to what the autopsis at Hamidia Hospital revealed. Rats, exposed for 4 hours to a concentration of 5 ppm suffered injury to the lungs and pulmonary edema. A cotton plug, saturated in the liquid for 30 minutes and applied to a rabbit, caused erythema, edema, necrosis and perforation. A few drops in its destroyed the tissues.

The course of MIC complications was divided into five stages by a German toxicologist:

1. The first stage, on the first day of exposure would produce acute and unbearably irritation of the eyes and the skin.

2+3, These symptoms would continue into the second and third stages, lasting from 3 or 4 days to a week, along with irritation of the lungs.

4. The fourth stage, developing over 3 or 4 weeks, would be characterised by acute systems of malfunctioning of the central nervous system (CNS).
5. The fifth stage would show up delayed symptoms of CNS disorder, including paralysis and other complications within five days of the gas leak, the first indication of the delayed effects had already begun to appear.

On December 5, 1984 itself 11 cases of cerebral paralysis were admitted to the Hamidia hospital. Cases of corneal ulcer with the possibility of corneal opacity were widespread, as well as the gradual development of pulmonary edema. In a bid to prevent secondary infections, like tuberculosis or pneumonia, the victims were being injected with omiticillin and tetracyclin, but no antidote. In Hamidia hospital's paediatric Ward, three children died of failure of central nervous systems—from intense convulsions and coma.

On December 9, 1984 a medical student volunteer, one of the hundreds engaged in relief work, was admitted to the hospital with serious motion disability. It was not until a German toxicologist, Dr. Daunderer arrived, that thiosulphate was administered to the gas victims on December 8, 1984. Daunderer's laboratory assistant, Karin Gottschalk, also performed the Acid test among others on the blood samples, and according to well informed sources, found cyanide in several cases.

What is now totally clear is that MIC does produce long term effects, including paralysis, even if it is still not clear how it breaks down, and whether the antidote is thiosulphate. This slmply means that the death toll will continue to rise month after month and year after year. It is also now obvious that the majority of the deaths occurred not on the first or second day from acute exposure but from developing complications. The majority of the dead were children and the aged. Yet, over the entire period, the treatment as largely symptomatic, with thiosulphate having been rejected without even being tried, or an alternative proposed. The thiosulphate antidote was administered in over a 100 cases with promising results. In several, the improvement was spectacular, and acutely affected patients were discharged in two days. Although the antidote is available in India, and 200 cases were called for from Indore, Daunderer had to ask for them to be flown in from abroad. Without an antidote, there is nothing to stop the victims from slowly dying, or being afflicted with blindness and secondary complications like tuberculosis.

It was at 1.05 a.m. on December 3, 1984 the superintendent of police of Bhopal was first alerted to the leak by a telephone call at his residence by the town Inspector at Bajaria. People were fleeing the town in their night clothes, on foot, their children in arms. Within fifteen minutes the SP of Bhopal was at the police central room. First, there was a chill in the eyes, in the nose. Then it was like a sharp point of a knife being drawn down your gullet. Suddenly you felt suffocated. The police control room was flooded with killer gas.

According to a report in the *Illustrated Weekly of India* (December 30, 1984), the first act of the SP was to call the Union Carbide plant officials. Whoever picked up the phone could not tell him what was happening. The SP, Mr. Swaraj Puri asked to be put on to a technical man, anyone in charge he could, but they couldn't be able to say anything. He asked for the plant manager to report to him immediately, he assigned one of his men the task of calling Union Carbide incessantly. It was getting difficult to breathe, their eyes were burning and watering. On an instinct, Mr. Puri rushed into the adjacent rest room, and splashed his eyes with water, then turned his face up to the fan. He felt some relief. For the next hour and a half, they rushed between the water, the fan and the phones. It was 1.30 a.m. but there was no information at all from Union Carbide. One by one, his men began to leave the control room, unable to take the killer gas any longer. Between 1.30 and 1.45 a.m., the Additional District Magistrate called Mr. Mukund, Union Carbide Manager at his residence: "How can there be a leak at my plant?" Mukund said, "It had been shut down."

Most people in Bhopal knew there was nowhere else it could come from, but Union Carbide. Reports were coming in on the wireless and on the phones, of people thronging the streets in thousands, fleeing. The situation had to be controlled somehow. It occurred to Mr. Puri that anti-social elements could begin to exploit the situation, looting the houses that people had abandoned with their doors open. The panic had to be checked. Accordingly, he issued instructions to his wireless vans to announce that there was no need to panic, that they must not leave their houses.

"If I had only known from Union Carbide that a poisonous gas had leaked," Mr. Puri said, "I would have issued exactly the opposite instructions—to immediately evacuate."

A sennior plant manager off duty that night, called the plant from his residence, but couldn't elicit any information either. A security consultant to Union Carbide, a former IGP himself, called the control room to ask what had happened. The phones rang inceassantly. There were calls from senior government officials, as well as from the ADC to the Governor, asking what they should do. Mr. Puri suggested to use wet towels on the face, without having the slightest idea how deadly that poisonous gas was.

Mr. Puri by then was the last man left in the control room. A few minutes earlier, he had ordered KPS Chauhan, the last of his men, to leave the control room. An old man, Chauhan was in such a bad shape that he couldn't take off his uniform. Between 2 and 2.15 a.m. Mr. Puri first heard the siren. It was unmistakable and could be heard all over the town. It was Union Carbide's siren, but still there was no word from the plant manager. Between 2.45 and 3.00 a.m., a plant official who introduced himself as Mr. Mehra, came to the control room to confirm that "the leak had been controlled."

"That was obvious", Mr. Puri told him.

"I can see there is much less of it in the air now, Mr. Puri affirmed once again.

Mr. Mehra was the first official to conform that there had been a leak of poisonous gas from Union Carbide plant. But the disaster was complete by then.

As it turned out by then, there was no leak left to control—tank No 610, which held 40 tonnes of MIC, was now empty. At first light of the dawn, Mr. Puri drove down to the site.

Across the road from the plant was Jayaprakash Nagar, a huge slum colony that ran almost the entire length of Union Carbide's boundary wall. He found only corpses and corpses. The havoc was wrought within a 40 square km area. Inducing sensations of burning of the eyes, hoarseness in the throat and finally death by asphydiation on a sleeping populace. It was a scenario straight out of any disaster formula movie. Only difference was that it was real and scenes were cruel beyond words. As slumbering citizens of Bhopal reeled under the effects of the killer gas around 1.00 a.m. on the morning of December 3, 1984, husbands woke to see their dead wives by their bedside; monthers got up suddenly to find their children close to them either dead or are dying, fathers woke up to see that the entire family was gassed to death. The survivors in panic took the deserted streets which all of a sudden became alive to the exodus of thousands of people moving, walking and running away from death. They were moving out of the poisoned city. To the railway station, to the bus terminals, into the hill top, into the open spaces, anywhere but for away from Bhopal—a death city.

The avenues of escape were not open. At the railway station, the crowds found that all rail traffic to and from the city had already been stopped. The railway station master and the booking clerks were among the early victims of the killer gas. Soon the effects of exposure to the 'thick white cloud' of MIC—began to sink in on even the fleeing population. And precisely, the first victims among the long march were children. Crying out as their eyes burnt and their throats were consumed by a raging fire. Until they collapsed into eternal silence in the arms of their parents. The adults fared no better. They sank to their knees and lay down on the streets. Gasping for breath. Most of them were left to die in agony. A few were rushed to the hospitals where doctors, peplexed at the symptoms could not do more than administering first aid. And in effect, leaving several of the early victims to their fate, soon people were falling like flies dead on roads. And also in the hospitals, especially Hamidia Hospital, where doctors had to scour the laws to segregate the victims who had died and the victims awaiting treatment, from those who could still be treated. It was not yet dawn. Yet human agony was unabated.

When the Sun rose that day, Bhopal city was turned into a city of death, if not desolation. On the streets were hundreds of dead bodies of people who had given up the flight to a seeming escape from the killer gas. Overcome by the deadly fumes and frightened by the run for life, the entire family one after another collapsed and perished together. Days later several survivors recalled memorable scenes of that traumatic night.

People were trampling over bodies of the dead and dying in a mad rush to stay alive among the dead, or running past their dying neighbours, friends, and sometimes even close and dear ralations in the frenzy to stay ahead. And also, sometime, entire families were sinking to their knees, sitting on the roads and waiting for death:

"We lived together and we will die together", was their refrain. Though the official figures had already put the dead body count at over 2,500 and several thousands more must have perished in the tragedy outside the hospital records. By the estimation of the state government itself, no less than 25 per cent of the city's population of 8 lakhs was affected when the deadly poison gas escaped into the atmosphere on the night of December 2 and in the early morning of December 3, 1984 from the underground tank at the Union Carbide pesticide plant at Bhopal.

COMPARISON BETWEEN BHOPAL MINAMATA TRAGEDIES

Raj Kumar Keswani draws parallels between the Bhopal and Minamata tragedies (*The Illustrated Weekly of India*, April 27, 1986). The advanced technology in both the cases has caused these ttagedies While 30 years age, it took several years to kill a few hundred people in Minamata, it took a few hours to kill several thousands in Bhopal. Growing industrialisation has made one indisputable advancement; it has learnt how to kill at a much faster speed than ever before. The people of

Minamata and Bhopal have been the victims of industrial pollution; comparisons between the twe tragedies are, therefore, inevitable, and raise more disturbing questions. Until the Bhopal catastrophe siezed the public consciousness, the haunting images of the Minamata tragedy symbolised the genocide industrialisation inflicted on nature and on man. Through the late fifties and early sixties, hundreds of inhabitants of the Japanese town of Minamata wasted away. The victims of a debilitating disease of the nervous system caused by mercury poisoning were transformed into a dumb, helpless creature. For these victims, it must have seemed as if the world was fading away from them. Paralysed, their hands suddenly could not hold anything; their eyes could not see anything and their ears could not hear anything. Out of fear they cried aloud but their voice would not come out. This strange disease was later identified as the Minamata disease.

Minamata is a small town with a population of approximately 30,000 situated in Kuamoto prefexture in the south-west part of Kyushu Island in Japan. In Bhopal the culprit was the US multinational Union Carbide and in the case of Japan the culprit was the Chisso Corporation, which was also a multinational of sorts in its heyday. The word 'Chisso' means nitrogen and as the name indicates, the company was engaged in the production of nitrogen fertilisers. Later, the company started using inorganic mercury as a catalyst for manufacturing acetydehyde gas. A byproduct of this process, methyl mercury, was discharged as waste into Minamata Bay. This industrial waste along with other compounds, entered the food chain through fish and shellfish in the Bay.

Minamata being a fishing and farming village, the bay fish constituted the most common food of a majority of the inhabitants of the area. The first human victim of the disease entered the Chisso factory hospital as far as April 1956, The government took 12 years to officially declare the cause of the disease and finally, on September 26, 1968, the government announced that Chisso Corporation was the cause of the Minamata disease. Almost four years before the disease was first identified in human beings, cats started committing suicide.

Cats suffered from severe mercury poisoning which was concentrated heavily in the internal organ of fish. After consumig the poisoned fish from the Bay cats began killing themselves by jumping into the sea or into the kitchen fire. After the cats came the human beings. These victims suffered from numbness of hands and feet, heavyness and dullness, pain (of head and shoulders), heaviness of the head. The victims could not move finger and dropped articles from hands. The disease was caused due to heavy metal type of poisoning as a result of eating fish and shellfish of Minamata Bay which was polluted by the industrial wastes known as methyl mercury.

According to the report of the Board of Science and Technology methyl mercury compounds were discharged from the factory of Showa Denko into the Agane river for a long time, and they had a connection with the occurrence of poisoning, and were the basis of the affair. It is impossible to decide whether the pollution resulted only from this long term pollution or if there was another short term pollution in addition to the original one. There are two possibilities for a short term pollution; one is inadequate control of waste before the closure of the factory in January 1965; the other is a possible discharge or dumping of agricultural chemicals stored around Niiagata harbour at the time of the Niiagata earthquake. It is impossible to decide which was the cause of the short term pollution because of lack of sufficient information at present. What happened in Minamata in a span of 30 days by man-made pollution, is not curable even after 30 years.

To begin with, the similarities between the two disasters. In Bhopal, even after 16 months no anidote or effective treatment has been found for the gas victims to date. In Minamata, even after a lapse of 30 years, no effective medication has been found for the gas victims. In Minamata, it was the cover-up tactics of the Chisso Corporation which prevented the exposure of the real cause of the disease. Exactly, the same strategy was adopted by the Union Carbide in Bhopal with the result that even today it is not known exactly what the "other gases" are which are still causing the people to suffer.

The Chisso Corporation in Minamata created its own group of scientists to disprove the findings of the research group which came in the conclusion that the strange disease was caused by methyl mercury. Union Carbide followed exactly the same line of action and successfully introduced a number of scientists who did nothing but add to the prevailing confusion. May be, years later, the truth might surface that these 'famous scientists' were indeed sponsored by Carbide as was the case in Minamata where it was proved years later that the scientists who supported the interests of Chisso were adopted by the company.

The attitude of the government in general and the bureaucracy in particular was no different at both places. While in Minamata, the bureaucracy helped Chisso forge the papers and warned the agitators to accept the settlement offered by the company, the Bhopal victims were forced to sign the authority in the favour of the government in the matter of compensation. But Bhopal went a step further. At the right time, right information which was at times exclusive, was fed to the Carbide management by a few overzealous bureaucrats. Sometimes this information was not even made available to the state's chief minister. The bureaucracy acted in the true spirit of 'private servants'.

In the case of Minamata, the worst affected belonged to the lowest strata of society; mostly poor fishermen. In Bhopal, a majority of those who suffered were either labourers or daily wage earners. While in Minamata, the victims had to face the wrath of a section of society, Bhopal was no different. In Minamata, a section of the people condemned the movement of the victims because they were convinced that their survival depended solely on Chisso's survival. In Bhopal, the big shots of the society were indeed jealous of the jhuggiwale who were getting all the attention, free rations and financial assistance.

Difficulties in identifying the victims of both the disasters also show glaring similarities. The occurrence of a few symptoms of the Minamata Disease was denied on the pretext that the strange disease can not be caused by methyl mercury alone. In Bhopal, the symptoms were attributed to smoking and tobacco chewing. Both the tragedies brought to the fore the "couldn't care less" attitude of political parties and exposed their "real concern" for the welfare of the people. In Minamata, the political parties adopted an attitude of indifference and refrained from active participation in the victims' struggle in the initial stages. The case of Bhopal was similar: no leader worth the name of any national-level party found time to come and visit the affected areas.

Equally significant are the contrasts between the two disasters. While the movement of Bhopal victims calmed down within a year of its commencement, the Minamata struggle is a continuous one and going at full force for the last 30 years or so. A glance at the Bhopal movement would make it evident that no group was able to create or leave an impact except for getting some media coverage, whereas the Minamata movement was started and sustained by the victims themselves.

It was at a later stage that they were able to muster the support of activists. But the fact remains that the victims were always in command and the reins of the movement were totally in their hands. In Bhopal, the activists were unknown to the victims with the result that the activists failed to get at the real core of the victims' problems. The government took advantage of the situation and even labelled them as CIA agents. Later when the groups split, they played against each other. This made the already confused situation more confusing and the real issues never emerged. In Minamata, the movement was directly aimed against the Chisso Corporation, which, in the eyes of the victims, was guilty. In Bhopal, the wrath of the victims got diverted towards the government, thanks to meddling bureaucrats and a few confused leaders. This is how Raj Kumar Keswani has narrated a tale of two cities.

2

BHOPAL TURNED INTO A GHOST CITY

AFTER THE TRAGEDY

THE FAILURE, of the safety systems at the Union Carbide pesticides plant in Bhopal was confirmed by Dr. S. Varadarajan, Director-General of the Council of Scientific and Industrial Research, who told the Science Congress session that a multi-disciplinary effort for the study of the impact of the gas tragedy would be continued. Dr. Varadarajan also highlighted the dangers and uncertainties involved in the operation to neutralise the remaining methyl isocyanate in the plant thus contesting the view held by some that it was a routine operation and the authorities had unnecessarily created panic by taking elaborate measures, including the shifting of population from the vulnerable areas.

Dr. Varadarajan also stressed the importance of basic research and by way of illustration pointed out that some basic chemistry calculations were required before undertaking the disposal of the remaining dangerous material in the plant. The operators of the plant, who had to be assembled with great difficulty, were of little use in this regard since they had been accustomed to only the given manuals.

Dr. Varadarajan who had co-ordinated the "Operation Faith" after the tragedy, recalled that the man-made accident caught the community, the government, the scientists and the medical profession unawares. The affected people did not know what preventive and remedial measures could they take when the fatal gas started leaking out and engulfed human kind to disaster and death.

In the resulting panic and chaos, the people did the wrong things and soon there was no wood to burn the bodies and no land for burial. However, the combined efforts of the government, scientists and doctors ensured that there was no outbreak of communicable diseases after the gas tragedy.

From the account of the accident given by Dr. Varadarajan, it was clear that all was not well with the plant. He referred for the broken safety valve, the level controller in the MIC tank not working and the alkaline scrubber for detoxification of MIC unable to deal with the situation in which MIC liquid turned into gas. There was no efficient cooling system. The pumping cooling system that was there was hardly adequate.

Dr. Varadarajan also could not figure out how one and half tonnes of water got into the system unless there was a big leak. After the tragedy, the disposal of 15 to 20 tonnes of remaining material was a big challenge. There was a question mark about the stability of the stocked material. There were problems related to processing for detoxification and containment of gas leakage, if any. It could not be ascertained whether the stored material was stable or not.

All possible alternatives were examined and precautions considered. To rule out any disturbance in the inputs, the entire input-output system was changed. The vital instruments were replaced with zero defect instruments. Experimental stimulation was done for confirming the rate of reaction. It was also considered that no liquid may come out if the material had already been polymerised. Among the other steps taken were the folding of the outlet. The emission tower was blocked. The area was cordoned off by a wet blanket and arrangements were made for aerial spray of water. Ironically, while it was water that leaked into the system which led to the reaction and to the accident, it is water that is an antidote for those who suffer from MIC poisoning. Unfortunately, not many victims knew about it. Those who remembered to inhale through a wet towel were able to minimise their suffering and the ill-effects of the gas.

Dr. Varadarajan said that one-third of the 45 tonnes of MIC in the tank turned into a kind of plastic (polymerised) while the rest became gas and escaped resulting in the tragedy. He said that phosgene which is normally added to MIC to "inhabit" polymerisation might have actually initiated the disastrous reaction after entry of less than 500 grams of water. The small amount of water was enough to react with phosgene liberating hydrochloric acid which easily polymerises the MIC. As enormous heat is released during the polymerisation, the rest of the MIC became gas had escaped after rupturing the safety valve. Dr Varadarajan made it clear that water reacting with phosgene was one of the two reasons for the accident.

"Operation Faith" to dispose of the deadly methyl isocyanate (MIC) in the US multinational Union Carbide plant ended with the conversion of the entire stock of nearly 23.5 tonnes of MIC into pesticides carboryl (sevin) on the seventh day, according to official sources. The MIC disposed of included 22 tonnes in the underground storage tank and pipeline and nearly 1.5 tonnes stored in the six stainless steel tanks.

The conversion work which started on December 16, 1984 was accomplished by the Carbide staff under the supervision of a team of top Indian scientists, headed by Dr. S. Varadarajan, director-general of the Council of Scientific and Industrial Research (CSIR).

The plan to utilise the MIC still remaining in the US multinational Union Carbide pesticide plant suffered a setback with the discovery that the quantity left exceded the original estimate by a wide margin. This was confirmed by Dr. Varadarajan, who headed the team of the Indian scientists and engineers working in close collaboration with the US multinational executives in implementing "Operation Faith."

According to Dr. Varadarajan, the US multinational plant contained net 15 tonnes of MIC when the conversion operation began. By December 20, 1984, 18 tonnes of this total gas had been converted into pesticide carboryl (sevin). This quantity was drawn from MIC storage tank number 611, and excluded the amount that might be contained in tank number 619 (supposedly empty, but according to Dr. Varadarajan, not necessarily so), as well as in six stainless steel tanks, which were estimated to contain 1.2 tonnes of MIC. Tank number 610 which had discharged its load of about 40 tonnes of MIC into the atmosphere via the scrubbers, was free of the toxic chemical, although some polymers of the substance and impurities were there.

At a press conference on December 20, 1984, at Bhopal Dr. Varadarajan admitted, for the first time, that the effective control and supervision of the MIC conversion operation was vested in the hands of the executives of the US multinational. This statement made it clear that his team of experts had at best a peripheral role in the "Operation Faith." He also mentioned that besides the relatively pure MIC contained in the storage tanks, there was systems. This was likely to be of low quality and unfit for conversion into carboryl. It was neutralised under controlled conditions in the caustic soda scrubbers. The reaction was likely to excude an unpleasant odour. However, he added that Bhopal was "completely safe" and that there was no cause for worry.

Even as the issue of disposal of MIC attracted increasing public attention and anxiety, the team of scientific and technical experts from various Indian institutions who gathered at Bhopal seemed to be badly divided over the method to be employed to detoxify and neutralise the stock of MIC still lying in a storage tank at the plant. Three entirely different methods had been tentatively proposed for the safe disposal of the toxic chemical. One was to react it slowly with chemicals that would neutralise it to produce harmless substances. The second method would involve incinerating MIC under pressure and controlled conditions. And the third would be to use MIC as an intermediate in the US multinational's plant to produce carboryl.

The last proposal had been just made and been vigorously pursued by Union Carbide executives. It apparently found favour with some senior scientocrats of the directorate-general of technical development and the CSIR.

The second proposal found backers desperate among a group of scientists but particularly chemists who theorise that the toxic releases, such as amines, from MIC combustion would not be dangerous beyond "acceptable" or "reasonable" limits (which remained undefined and subjective). However, the safety of this method had not been established even tentatively.

The idea of gradually neutralising MIC with chemicals right inside the plant and without transporting it to another site had won among environmentalists and toxicologists. Among the substances that were being mentioned was caustic soda, which was easily available and whose reaction with MIC was fairly well understood.

Chemical neutralisation could be a slow process but it was relatively safe and simple to handle, argued many young scientists and engineers. They pointed out to the experience that Vietnam had with detoxfying and decontaminating large tracts of land, many homes and factories which were bombered for years and to saturation levels with lethal herbicides—a part of American plan during the Indo-China war to defoliate the country.

The Vietnamese were considered to be the world's greatest experts in handling the problems of pesticide detoxification and de-contamination. And many of the scientists gathered at Bhopal asked why the Vietnamese or independent experts from Europe who had actual experience of detoxification were not being called in to assist their own team.

If the drawback of the chemical-detoxification proposal was that it was slow and dilatory, the principal problem with the plan to use MIC to produce carboryl in the US multinational plant was that the plant had itself proved, and with lethal expert, to be unsafe. There was indeed no guarantee that the carboryl plant could function satisfactorily or safely, said engineers supplying the first proposal. For it had already developed maintenance problems before the accident and if it did not run efficiently, the processing of a fairly sizeable quantity of MIC could itself take a very long time. Besides, it would tantamount to granting US multinational some sort of safety approval or certificate and record to letting its management, with its all-too visible safety record, ran the plant as it liked.

A minor and significant accident occured at the plant during the "Operation Faith." The mishap consisted in the overflow of a slurry of carbon tetrachloride and carboryl pesticides from a vessel known as "surge tank" in the sevin plant at Union Carbide in Bhopal, in which the MIC utilisation programme was underway. At least two persons including a Union Carbide official were exposed to the toxic mixtures. The surge tank overflowed when an additional quantity of the slurry, produced in the sevin reactor, was pumped into the vessel with a view to evacuating the reaction quietly so as to speed up the MIC conversion operations. Just before this happened, the surge tank was showing a high level of between 80 and 90 per cent of its capacity.

The backdrop to the event appeared to be as follows:

> Owing to a problem with the filteration process (the last stage of production of sevin, a solid insecticide, the surge tank could not be emptied on time. The tank was the vessel to which the contents of the sevin reactor were normally transferred once the cooking of the MIC with Napthol had taken place. Since the tank was almost full, the reactors's contents should not have been transferred to it. However, a certain amount to haste and impatience appeared to have marked the decision to do so. This was rooted in the effort to get the reactor empty as fast as possible so that it got ready to receive MIC from the storage tank Number 611 via the charge-pot. This haste appeared to be linked to the attempt to meet the deadline for the completion of the MIC conversion job. The consequence of the overflow was that at least two Carbide employees had the toxic slurry falling or splashing over them.

Both the chemicals contained in the slurry were toxic. Carboryl was a pesticide which was poisonous by ingestion, inhalation and skin absorption. It was a bad irritant, its threshold limit value was only five miligramme per cubmic metre of air for 8-hour working day. Carbon tetrachloride irritated the eyes, the nose and throat. It was narcotic in its effect and sensitised the skin producing dermatitis on prolonged exposure. Its vapour could cause dizziness, nausea and liver or kidney damage. In acute cases, it could produce vomitting and stomach pain, unconsciousness and inflamation of lungs. The threshold limit value for the chemical was only 10 parts per million. It was not known if the Carbide employees were badly affected and what their condition was like. This mishap had apparently not been the only abnormal occurance since "Operation Faith" began.

The Central Bureau of Investigation (CBI) officials investigating the gas tragedy got to work on the MIC as tank at the plant. One foot square hole was made in the tank at the plant to remove some polymerised residue. A few more holes were made in various parts of the 40 feet long MIC storage tank as part of the residue collection exercise carried out under the supervision of six scientists. The operations continued to mark the final stage of the CBI investigation.

Informed sources said the purpose of the exercise was to ascertain the quantum of the residue material. The CBI team was already in possession of the results of an analysis of the residue samples removed earlier from the tank. Information regarding the weight and quantity of the total polymerised material scrapped out of the tank enabled the analysis to assess the scope and extent of the chemical reaction that had caused the catastrophe. It helped to determine the quantity of MIC gas that escaped into the atmosphere. The opening up of the tank enabled the team to find out whether any corrosion under excessive heat had occured in the tank, made of non-corrosive material.

The inquiry report said among other things, that a sizable quantity of water (between 450 litres and 900 litres) introduced into the MIC tank "inadvertently or deliberately" had caused the catastrophic chemical reaction. The Central Bureau of Investigation (CBI) officials had also had a similar experience. A few days before the start of "Operation Faith", the indicator of the tank in which 15 tonnes of MIC was stored, showed an increase in pressure. It was discovered that the sodium hydroxide flushed into the vent gas scrubber had filtered down up to the rupture disc.

The American team of the multinational suggested certain measures to prevent the entry of the foreign material into the tank, but the CBI staff on duty vetoed it. The matter eventually went up to Dr. Varadarajan who gave the Americans the go signal. When asked by newsmen about the occurance of any mishap after December 3, 1984, the CSIR chief merely said: "not that I know of".

A great deal of confusion prevailed over the precise role that the team of scientific experts drawn from various government laboratories and institutions of the country and headed by Dr. Varadarajan of the CSIR, was meant to be playing in the MIC conversion operation at the Union Carbide plant. This confusion also rested on the root of the uncertainty that existed over the steps to be taken after the job of MIC conversion was completed.

The official position was that Dr. Varadarajan was in overall charge of the plan of utilising the MIC remaining in the plant to make carboryl pesticides. Dr. Varadarajan had himself been at pains to emphasise that he was actively supervising the operation and even administering the multinational's management in its conduct. In fact, the US multinational personnel operated the system under the supervision of the Indian team of experts. Among the multinational's staff—with whom Dr. Varadarajan admitted to have held consultation—was a five member committee of American multinational including Woomer, a former works manager of the Bhopal pesticide plant, the local works manager, J. Mukund, who was the son of a former governor of the Reserve Bank of India and the relative of an influential diplomat, also played an active role in the operation along with UCIL's four other Indian employees arrested after the mishap. Converting the killer gas was the American show and they carried out the task by running the two 12 hour Shifts (unlike the normal three eight hours shifts) in which the working of the plant was divided.

Not a single Indian scientist or engineer or any other expert from the government had actively worked inside the operating area of the plant during the period as a matter of faith. According to the plant's active personnel, the experts' work was confined by and large to be small room on the ground floor of the administration building, which itself was about 200 metres or further from the carboryl production facility. The room had been equipped with nothing other than a telephone and a walkie-talkie set, by means of which the Union Carbide operators periodically communicated to the experts some information on the quality of MIC being transferred from the storage tank to the charge in the sevin unit.

The Indian Government experts played no role in checking the safety systems, valves or even pressure level and temperature gauges. There appeared to be no truth in the report that Dr. Varadarajan or his associates advised the US multinational's engineers and actually supervised the operation. In fact, Mr. Woomer had been quoted as saying that he advised Dr. Varadarajan and not the other way round.

There was indeed some sort of understanding that had been reached between the US multinational's managers and the government scientists. The understanding was about leaving the entire operation of MIC conversions, with all its details, to the company's engineers and putting the latter in charge of the plant as a whole, and conversely letting the government experts devise the so-called "addition safety system"—such as mounting jute sacking on the perimeter wall, wetting the outside roads with water, etc.—all essentially outside the purview of the core of the operations and peripheral to its progress. A component of this understanding was that in public, Dr. Varadarajan would speak for the operation as if he truly supervised their job.

That Mr. Woomer, not the Indian scientists, carried out the operations was amply proved by entries in one of the maintenance cards newsman read inside the control room of the pesticide plant. The card relating to tank number 611 (in which 15 tonnes of MIC was stored), read:

> "Reasons for outrage: requires approval by CSIR for operations as per Mr. Woomer". This fact, coupled with the experience of the CBI, conveyed the whole story. Dr. Varadarajan was merely a front-desk man. The Americans were the ones who actually carried out the "Operation Faith".

So the truth was that the Indian team of experts played no role in checking or measuring the various parameters, including the quantity of the MIC present, leave alone in operating the plant. They were wholly dependent even for their information on the US multinational's employees, as the contradictory statements made by Dr. Varadarajan on the quantity of MIC remaining in the storage, showed.

According to the Times of India News Service of December 20, 1984, the understanding, in one sense, itself reflected in the camaraderie now much in evidence between the two sets of officials—was traceable to the decision made in favour of converting. (as distinct from neutralising), MIC into carboryl. This involved the rejection of the other options of dealing with MIC discussed amongst the scientists and experts and the handling over of the operation to the US multinational management. It meant using the term "neutralisation" (strictly rendering safe and insignificantly toxic) to describe the processes of making a pesticide, itself a toxic chemical, and also implied an endorsement of the Union Carbide process of conversion as safe and of very detail of the operation as essentially sound. The last element included the vent gas scrubbers which failed on December 2/3, 1984 and which in any case, was incapable of neutralising MIC should a big release occur.

The second most significant aspect of the retreat of the Indian experts from the nuts-and-bolts level of the operation related to their lack of familiarity with the plant, its equipment, process and control system as well as its routine operating and maintenance procedure. Most of the experts who had gathered (and many of them had left) at Bhopal under the leadership of Dr. Varadarajan had been pure scientists or engineer-managers rather than nuts-and-bolts engineers dealing with the details of running an industrial plant.

Thirdly, most of them were not even familiar with equipment used outside laboratories and pilot plant (or other industrial factories) and therefore could not effectively intervene in the supervision of the operation. As any rate, the result of the operation had been a net gain for the multinational, which had been allowed to recoup some of its heavy losses. On the other hand, many scientists complained that they had not been encouraged to do the kind of work that they were best equipped to do in Bhopal. This included mapping the dispersal of the plume of MIC that had been released from the plant, determining and quantifying the extent of actual health damage done, measuring with a high degree of accuracy the presence of toxic substances (products of the gas upon reaction with water, bodies, air, vegetation, etc) in the environment.

In fact, no reliable data, itself a valuable asset for understanding the short and long term of the gas leak had been generated by the experts at that stage. Such tests as had been conducted on vegetable samples by a research institute were done in a sloppy manner, with unstandardised methods without specifying the objects of the analysis and using unspecified instruments whose accuracy was not known, said scientists.

The results of the analysis had, expectedly, failed to assure the public that there had indeed been no damage to the environment and that the air, water, vegetables, meat and poultry of Bhopal were safe. Such assurances were, at any rate, held out even before the results were available. Meanwhile, environmentalists and new groups of them had sprung up at Bhopal—began to collect samples on a systematic basis and to get them analysed in reliable laboratories in the metropolitan cities of India.

The Bhopal water was well a factor in sustaining a high plasma thiocyanate level in the blood of the gas hit. This was indicated by a medical survey conducted by a team of specialists and laboratory technologists. Thiocyanate was the form in which cyanide was found in the human body. According to the report of the survey team released to the press in May 12, 1985, the average thiocyanate level in the blood of the Bhopal gas victims was 2.6 ppm (part per million) against 0.9 ppm in respect of the control group in Bombay examined by the medical team for the purpose of comparison.

The thiocyanate level in the water of the upper lake, main source of water supply to Bhopal, was ppm. It was 3 ppm in the lower lake and 2 ppm in respect of the filtered water. The thiocyanate level in the tap water in Bombay was found to be 0.8 ppm. The water samples were tested in March, 1985.

The survey team, comprising 21 medical specialists and laboratory technologists, was sponsored by the Citizens Relief and Rehabilitation Committee in Collaboration with the Voluntary Health Association of India in New Delhi and the Bhopal Relief Trust in Bombay. The survey team examined a total of 569 gas victims forming a representative sample of the people in the gas affected localities.

The survey-findings regarding the thiocyanate level in Bhopal water would have been more meaningful if corresponding figures were available for the period before December 2/3, 1984 gas disaster. However, it was suggested that the public health authorities should follow up the survey findings with further test of water sample to assess the change in the thiocyanate level in water and also take effective steps for decontamination of it.

Another according to the survey, continued exposure to "unusually high levels" of thiocynates, and detected in the Bhopal water can affect the functioning of thyroid glands. High thiocyanate level in blood is known to be one of the causes of goitre.

Another notable finding of the medical survey pertains to women, is the 198 women examined by the team, 100 had gynaecological symptoms. The main complaint was about "excessive vaginal discharge". According to the survey report, 29 of the 38 pregnant women in December, 1984 reported abortion in their first or second trimester. A common feature in 17 pregnant women examined by the medical team was anaemia. The women also reported "reduced foetal movements." There have since been two cases of birth deformed babies. The deformities were similar in nature.

Although a spokesman of the survey team briefing newsmen said it could not be conclusively stated that the cases of deformity were the result of MIC exposure, the survey report suggested that the pregnant among the gas affected should be made aware of the possibilities of abnormal birth and also given proper guidance regarding continuation or termination of pregnancy.

The survey team is reported to have found in the gas affected women that the growth of the foetus in the uterus was not normal. "This leads us to believe that there are some agents present in the body which are causing abnormality in the growth of the foetus, the report said.

As for the children covered under the medical survey, the report stated that persistent disabilities appeared to be much more marked in the children than on the adult examined by the team. Among the surveyed children, living within a distance of 1.5 to 2 km from the Union Carbide factory, the respiratory system was found to be affected in 50 per cent cases, the X-ray findings indicated abnormal condition in 67 per cent cases. The most striking observation was the "mental apprehensiveness" of the children.

OPERATION FAITH

For Operation Faith, the stage was set. The material inside the tank was "stable" and pressure and temperature were within safe limits. There was no fear of a further leakage but no exposure had been spared to adopt a chain of safety measures. The MIC was converted into carboryl, an agricultural pesticide by mixing it with Alpha Napthal. Under nitrogen presence it was pushed into the carboryl tank and this process of transfer was conducted only during day time to minimise risks. About three tonne of MIC was scheduled to be converted daily and about 3 to 4 days were expected to be required to convert 15 tonnes of MIC which was reportedly lying in the tank in a solid form. Actually it took a week to convert 18 tonnes of MIC gas into carboryl.

Dr. Varadarajan led the "Operation Faith" on December 16, 1984. The job was completed by the Union Carbide personnel under his supervision, so to say on December 22, 1984. The safety arrangements made were really massive involving technocrats from the US multinational and state government specialists and scientists from numerous fields and the defence forces. Indian Air Force helicopters were standing by to douse parts of the plant with water from above. Toxicologists, medical scientists, meteorologists, and men from various disciplines were there to study the operation and the effects of the lethal gas on ecology. The disaster underscored to such problems.

Indian Airforce helicopters, transport aircraft and highly specialised equipment arrived in Bhopal to deal with any exigencies during the "Operation Faith" at the plant. Three MI-8 helicopters fitted with special water tank had been kept in readiness at a nearly base to begin to douse water if anything went wrong. The IAF established a special control room near the plant which kept radio contact with the helicopter pilots and the experts working inside the plant when the neutralisation process started. The helicopters pilots wore special oxygen masks as were used by the fighter pilots.

One helicopter at a time continuosly hovered over the centre point of the plant at a height of 200 feet and doused water when operation began. IAF Dakotas which had been extensively used in mercy operations carried 120 oxygen-cylinders weighing 528 kilogrammes from Delhi to Bhopal. The Airforce also used its transport aircraft AN-325 and Avros for airlifting 110 doctors and nurses from Jabalpur, Mhow, Jhansi and Sagur to Bhopal.

The adequate number of army recovery vehicles had been provided to evacuate casualties from the plant site and nearby places to the army and civil hospitals and bring the patients home. Arrangements had been made on a war footing to tackle any situation that might arise during the neutralisation operations. Business and industrial activities in Bhopal came to a near halt and para-military forces took up positions on the city streets on the day of neutralisation.

Brisk preparations were apace at the pesticides plant to devise additional safety measures during the conversion of the MIC. The safety measures primarily consisted in the erection of a scaffolding of jute sacking round the carboryl (sevin) plant and the MIC storage tank, and dousing it with water to trap any gas leakage and reduce it to relatively less harmful chemical.

There was also a reserve of caustic soda available to neutralise any gas that might escape in case the water cover proved inadequate. However, it was not known if the storage tank would itself be kept cool or under refrigeration for the entire duration of the conversion operation.

There was a chilling and refrigeration system in the storage facility which could be employed to keep the tank cool and guard against a volatile reaction. It was not known, however, if the system was in working order. The jute sacking and the helicopter sprinkling water reserve treatment represented an additional line of defence against a potential MIC leak.

A preceding one, consisting of the scrubbers and the flare tower was reported to have been restored. It had not been possible to get an official confirmation of the report, because all information on the operation was centralised under Dr. Varadarajan. However, the flare tower was alight with the pilot flame.

Meanwhile, a shipment of alpha napthal arrived from Bombay. This was the chemical which MIC was finally reacted (after going through a minor stage of processing to yield what was termed "MRS make") in the carboryl plant to give the final pesticide product. With this shipment, rushed from Bombay, the pesticides formulation plant would have the quantity (roughly 30 tonnes) of alpha napthal necessary to react with the MIC stock remaining in the plant.

Fearing a new leak of the poisonous gas, thousands of people fled the city on December 12, 1984 following an announcement that a team of experts would begin neutralising the remaining 15-25 tonnes of MIC gas at the plant. Disregarding the assurances that there would be no danger, people thronged the railway station and the main bus stand where hundreds of buses had been mobilised from different districts. The announcement was made at a press conference by the chief minister, Mr. Arjun Singh, who said, if necessary, an estimated 1,25,000 people in the thickly populated areas around the plant would be evacuated.

After a full examination, the scientists came to the conclusion that since the gas leaked out of the plant while it was not working, the most practical safe way of neutralising the gas was the conversion into pesticide under the fullest safety measures.

The situation was confounded by rumours that the residential colonies around the plant were being evacuated. Police went round the city assuring the people that there was no danger and a similar assurance was broadcast over the local station of All India Radio. But there had been very little effect.

The exodus from Bhopal continued when thousands of panic-striken people left the town, ignoring the assurances repeatedly given by the state government that there was no room for anxiety. People wanted to be far away from Bhopal when neturalisation of the remaining 15-25 tonnes of MIC would be started in the US multinational's plant. According to a government spokesman, about 1,00,000 people already left Bhopal so far, but these were the people who went by train and buses only. Thousands more left Bhopal by taxis, scooters, trucks, vans and whatever form of transport they could get hold of, including tractors. Many families went out on foot carrying headloads and children. No account could be kept of them.

Without going into numerical assessment one could safely say that a sizeable segment of 2,00,000 people of Bhopal deserted the town, leaving behind them deserted streets, rows of locked houses and lerïe silence in places which usually buzzed with life. "Gas was the word that evoked fear in every heart and there was the stormy fear on every face Bhopal was turned out to be a town of unsmiling people, as a ghost city . . ."

Watching those waiting to board buses at the bus stand around mid-day on December 14, 1984 fleeing a town that was about to be bombed. Every outgoing vehicle was overcrowded with people on rooftop. Taxis and vans, the owners of which had chosen to stay on, charged sky high rates. A taxi ride to Indore cost Rs. 1000 against the normal rate of Rs. 250. But people were ready to pay out of fear of death. They were leaving in all directions—towards Indore, Gwalior, Ujjain, Jabalpur, Jalgaon, Rewa, Itarsi, Hoshangabad, Sehor, Khandwa, etc.

WHY PEOPLE FLED ?

In many cases the men were going to leave their women and children at the house of some relative in any nearby town and would come back to guard their houses and property against thieves who had a field day when the people had earlier fled their homes.

When asked why they wanted to leave their homes in Bhopal at all, when they knew that there was hardly any chance of the gas leaking again, most people answered that after they had seen such a holocaust they did not want to take the slightest risk for their families. They left Bhopal because they could no longer trust the words of the chief minister. When the top government officials sent their family away, how could common people believe what the administration said.

People were in no mood to believe the administration that had failed them once. They did not know what had happened on the night of December 2, 1984, when a lakh people convulsed with breathlessness in the three local hospitals. But they knew what was going to happen on the 15th December, 1984. They had little faith in "Operation Faith". Panic spread soon following rumours of another leak. The exodus began on December 12, the day 'Operation Faith' was announced. Crammed in trains, riding on bus stops, people fled the city to seek safety 200 km away. The state government's approach to the problem reflected its dilemma. On the one hand it stated that there was no danger to the city, and then covered the Carbide plant fence with thin canvas to be wetted constantly to render the MIC harmless in case of another leak. It said there was no need to evacuate the city but lined the streets with 400 buses to take people away from around the plant complex to the 10 camps opened elsewhere in the city. This did not help in generating confidence among the people.

On December 16, only closed doors and closed shops were to be seen. Main roads were crawling with siren-wailing police jeeps and tanks spraying water. The drone of a military helicopter overhead spraying water around the factory was a bizarre change from everyday bustle of Bhopal. Only 60 of the 5000 autorickshaws piled the roads and the fare ranged from Rs. 30 to Rs. 50 an hour. Taxi fare had gone up to Rs. 1,000 a day. The sense of panic, in fact, was unwittingly aided by the government's contradictory actions and statements. It had closed down the factory on the 3rd and announced it would not be reopened. So it was not for a week. And then, the government sprung upon the Bhopal population that the factory would be put back into operation to use up the remaining gas. One day before the operation began, reporters taken inside the Carbide complex for a look around were aghast to see the level and pressure meters in the Carbide control room still out of order. They reported as much and people armed with local papers wondered whether the plant was safe for "Operation Faith". Even if it was, nothing concrete was done to allay such fears in the minds of people. And in their haste to soft-pedal the fears arising out of "Operation Faith" the authorities claimed, without any real conviction, that only 15 tonnes of MIC had to be neutralised and that this would take four days. This was not so actually.

Announcements were made over the radio that neutralisation of gas would not mean another gas leakage but there was no stopping the people. A typical deserter was Bashir Khan who waited for a bus with his family of ten members. A central government employee, he was going to a relative at Sultanpur. A resident of Changarh, one of the badly affected areas; he had not lost anyone through the recent gas walk which caused his eye trouble and affected to some members of his family as well:

> "We have to go, for who knows what is going to happen," Khan said. But Khan was one of the lucky ones, who could afford to spend money for his transportation from Bhopal for the whole family.

In old Bhopal, an area with a population of nearly two lakhs, death visited countless families since the fateful night of December 2/3, 1984. Many died within the house due to gas leakage and many died subsequently after suffering for days. There were families without earning members, mothers, who had lost several children, babies without parents. They had to stay in their residences. They could not move out. For them there was nothing more to lose in life. They lost everything—even the last hope of living. They were merely living dead

Grim tales were heard from the residents of J.P. Colony who had yet to leave Bhopal on account of neutralisation operation. Situated across the street from the US multinational Union Carbide plant, the colony took the first blast of the lethal gas. There was so much deaths all around them that they could not shed tears, said a man who lost both his children. Mahadev Prasad, who lost his brother, a nephew and a niece, recalled that the morning after the disaster scores of bodies lay on the road. People carried, vomitted, fell and died. He had also fallen unconscious after sometime and woken up in hospital much later.

Only a few families were still there in the colony, a congested buster of rickety tin in timber or bamboo shanties in which about 6,000 people lived. Most of these shanties were destroyed. Everybody went away with life or without life. All around there was a shadow of death and the valley of uncertainty. In the midst of fear and anxiety there was a sting of death in the polluted air. All the sighs and sobs of Bhopal were added to the loss of life or life without any hope. The misery of living crippled and daunted their very existence as human beings. When people lost faith, no explanation was adequate to compensate it.

So an unprecedented rush was witnessed at the Bhopal railway station and bus stand soon after the announcement of assurances by the chief minister. Making it clear that the "Operation Faith" would start at 9 a.m. on Sunday, December 16, 1984 and there was no threat of any kind during the operation, likely to be completed in "four to five" days, the chief minister announced the government had made arrangements to shift out about 1.20 lakhs of residents from the old city, "should they desire so", during the period. However, those who opted to stay were helped and given "special instructions to safety practices during the period. The government already summoned about 400 buses and large number of army vehicles to help those leaving the town as well. For those who had no where to go, camps were set up in New Bhopal area. Most of these camps were set up in different schools and colleges in New Bhopal localities. The state government had announced that all schools and colleges in the city would remain closed till December 24, 1984.

The chief minister declared:

> "If required, the civil authorities would take the military's help too in the gigantic task of evacuation to localities within a radius of 3 km from the US multinational plant. The railways had also been informed to cater to the expected rush in traffic. He said the sick and invalid residing in the old parts of Bhopal were given help. All residents of the hutments surrounding the US multinational Union Carbide plant required compulsory evacuation. A special attention was paid to slums and hutment dwellers of old Bhopal."

Meanwhile, the Union Carbide team of experts which had flown in from the US along with its chairman Warren Anderson, a week ago also advocated this method rather than neutralising the gas by passing it through a scrubber. The chief minister said all safety measures were adopted while converting the gas into pesticide. Water was sprayed all around the plant by a helicopter to render the killer gas escaping through the vent gass scrubbers or flare tower ineffective. A foul odour spread the plant when it went into operation on Sunday but this did not mean any indication of danger. The plant was operated during the day and not at night for safety measures, to begin with. But the working schedule was subsequently changed and timing as well. Teams of administrative and police officers patrolled the entire city day and night to keep a constant watch on the situation. Reinforcement was called in from nearby areas.

The chief minister announced he had taken a personal pledge to be present during the operation in the plant. Newsmen were permitted to visit the plant on December 15, 1984, a day earlier of the operation. The localities whose residents were assisted to shift, "if they desired so" were the following:

Chhola Road, Jay Prakash Nagar, Railway Station, Chandbad, Sindhi Colony, Pullinghar, Shahjenhawabad, Immigate, Jahangirbad, Panch Sheel Cinema, Moti Masjid,

Budhwara, Kathikhana, Library Mangalwara, the Hamidia road bus stand, pirgati, Ibrahimpura and Jumerati. This covered almost the entire old city of Bhopal. They were camped at the TT nagar stadium, Model Higher Secondary School Campaign School, Nutan Girls' College. Subhas Higher Secondary School, Women's Polytechnic, the playground of the Bharat Heavy Electricals, Maulana Azad College of Technology and Satyasai College in New Bhopal area.

Buses were arranged for those desirous of going to their camps. The authorities made arrangements to shift the animals and a separate camp was set up at Hatai Khada on the outskirts of the city for this purpose. The chief minister assured the people that they would keep them informed through pamphlets, newspapers and the radio. He appealed to them not to give credence to any rumour. He called upon the people to display grit, determination, faith, trust and courage when scientists and experts started taking steps to save all citizens of Bhopal.

Still large part of Bhopal appeared as though curfew had been imposed even as Dr. Varadarajan announced that fool-proof safety arrangements had been made for the operation to neutralise and convert gas into pesticide at the plant. Most of the residents of the old Bhopal area left by night of December 14, 1984. Some shop keepers hurried out and it was shutter-down all over the market places and business centres. The state government's credibility had never gone so low before. The "Operation Faith" was one of the most crucial and agonising decisions that had been required to make and it called for all the commitment the government could muster and the rationality that it could bring into play. However, events of the last few days had shaken the eternal faith in human heart which would be difficult to restore.

The man in the streets refused to listen to official assurances:

> "We have no confidence in the administration", said everyone. People kept asking what was the need to set up ten large camps for 50,000 people in safe areas if there was no risk involved on "Operation Faith." The administration's

argument was that it must arrange shelter for those who wanted it. But till late afternoon of December 14, 1984 not more than five thousand people had gone to those camps. There were people who had nowhere else to go.

The majority of people in Old Bhopal belonged to low income groups but whoever could think of a relative elsewhere went away. From Shahjenhawabad to Jahangirabad, streets were empty. Policemen stood at lonely crossings in the usually bustling localities like Budhwara, Etwara, Jumerati, Hatikhana and Chhola Road. There had been fewer desertions from the better looking New Bhopal area but life came to a standstill there as well.

The fashionable New Market looked forlorn. Most shops were closed and the few which were open had no customer. The cars plying on the streets carried government officials or press-people. After Dr. Varadarajan's press conference, two bus load of reporters and photographers were taken to the US multinational Union Carbide plant in Bhopal.

The 60 acre plant area was deserted except for a few engineers and workers preparing for "Operation Faith" on December 16, 1984. Dr. Varadarajan and senior officials of the ministry of Chemicals and Fertilisers guided the tour. Not all questions were answered to the satisfaction of the press.

Dr. Varadarajan said an enormous amount of work had been done in the past 12 days to ensure that the people of Bhopal were not endangered during the operation that began on December 16, 1984 and continued for 7 days. He made it clear that the original material was not a gas but a liquid which had a highly reactive nature. Made out of Phosgene and methylamine, it reacted to water alkali and many other substances and it yielded a product like urea when mixed with water in proper ratio, if boiled at a temperature of 39 degree Centigrade.

Regarding the process of getting rid of the lethal gas, the chief minister noted that the experts had come to the conclusion that "most practical and safer way" among detoxification options proposed was the conversion of the left over MIC into pesticide "under fullest safety conditions." While the experts' opinion on the detoxification process was being awaited by all concerned with an air of expectancy, a sense of fear gripped the people, particularly those in the old parts of the city. About 50,000 persons were believed to have fled the gas-ravaged town since the early hours of December 3 when the US multinational Union Carbide plant threw up a cloud of the killer gas.

The rate of exodus in the last one week had been functioning like fever chart, depending on the day-to-day intensity of scare wave caused by varied rumours pertaining to persistence of air pollution, possible after-effects of the gas spread and apprehensions about fresh gas leakage. Every third house was found locked in the colonies along Chhola Road and near the railway station. There was unusual rush at the railway reservation counters and long queues of vehicles at most of the petrol pump stations in the town.

According to PTI and UNI reports, old Bhopal which dates from the 11th century and was humming with activity till December 1, 1984, housing nearly half of the eight lakh population of this state capital, was turned into a "dead city on the night of 14, 1984 again after December 2/3, 1984 with almost all its inhabitant absent as the countdown to the officially claimed "zero-risk" operation MIC gas disposal began on the scheduled date.

More than 2,000 security personnel spread out in the city as part of the action plan. The authorities announced on the night of December 14, 1984 fourteen do's and don'ts for the public to act on in case a warning was sounded. The people had also been told to keep tuned to the Bhopal section All India Radio which suspended all normal programmes and broadcast instructions about the precautionary measures.

Under action plan, adequate precautionery measures on a war-footing were taken for the safety of the scientists and the chief minister who were inside the plant, as well as those living within a four kilometre radius of the plant. Anthorities divided the worst affected areas in the old city of Bhopal into eleven sectors to be constantly patrolled by wireless-fitted jeeps that kept in touch with a control room at the state headquarters round the clock.

Men of the special armed force (SAF), pulled out of the decoit-infested areas, had also been deployed in the city along with the CRPF to look after law and order. Meanwhile, the exodus, set off by the fear of another gas leak, continued for the fifth day. About to lakh people were estimated to have fled from Bhopal. The temporary relief camps set up by the government had attracted only about 5,000 people so far, one-tenth their capacity. Traffic was thin on roads and several hotels in old city had announced closure and asked their clients to shift elsewhere.

Dr. Varadarajan addressing a crowded press conference said "the operation could have started earlier but we wanted to be absolutely sure that the MIC in the tank is stable." Asked what happened during the accident on December 3, 1984, he said it was caused either by water entering the tank or due to polymerisation of the liquid MIC.

Polymerisation (a process that turns the MIC into a kind of solid plastic) is normally prevented by adding inhibitors, but "we do not know if this was the case," he said. Both entry of water and polymerisation generate enough heat to vapourise the MIC which in this accident escaped after rupturing the disc of the safety valves. Dr. Varadarajan said half of the MIC had escaped and the rest had turned into a polymerised solid which was inside the tank.

He said at least two tonnes of water must have entered into the tank to cause a violent reaction:

> "We do not know what caused the accident only investigation will show", he added. Whatever the cause, it happened because the temperature inside the tank exceeded 90 degrees centigrade and pressure rose above atmospheres. He said leak occured at 11 p.m. and the escaped gas hugged the round because of an inversion. The slow south winds spread the gas over the populated area. As the outside temperature was only 14 degree Centigrade the gas once again became liquid that showered the city.

Had the winds been blowing north, deaths would have been few as that side of the plant was barren land. Dr. Varadarajan said the operation faith would be done during day time when there were no inversions. Four stations in the localities south of plant monitored the weather hour and a three-tier system of "water barriers" was in place. Water was the best agent to prevent escape of MIC as it turned it into harmless dimethyl urea.

Explaining the standby system, he said it for any reason MIC was not used to produce the pesticide carboryl it would be diverted to a huge caustic soda scrubbers which would be let out through a vent. As a further safeguard, the area around the vent had been turned into a "shamiana" a tiny room with walls of cotton fabric continuously sprayed with water. Two more tonnes of MIC kept in drums would also be disposed of safety:

> "There will be no dangerous chemicals left in the factory by the time we complete our job. "He said in reply to a question on the presence of phosgene in the factory. He said phosgene was not stored in the plant. The remaining MIC would be consumed in batches in one tonne vessels at the rate of about three tonnes a day. MIC would continue to be produced in the country as it has several other uses. But we must review the entire safety system in factories", he said.

Despite the government circular banning leave, there was an appreciable fall in attendance in the state government offices on December 14 and 15, 1984. December 16, 1984 was a Sunday. Most of those who did not report for duty were junior employees or those residing in the old city, where 11 municipal wards had been declared affected by the gas leak. Senior officers however, stayed carrying out the relief work and manning the various sectors into which the whole of Bhopal had been divided after the tragedy.

Frenzied mobs forced their way even into reserved compartments on long-distance trains. Nobody dared to stop them as they perched themselves on the top of the carriages and even on the couplings. Indeed it was the biggest evacuation the country had experienced since the partition of the sub-continent in 1947 by the British imperialist. "Over lakhs left the city by train. . . and 90 per cent of them were without tickets."

The Madhya Pradesh High Court directed the state government not to completely destroy the killer gas methyl isocyanate (MIC) under the "Operation Faith" which began in the Union Carbide plant at Bhopal on December 16. The vacation judge, Mr. Justice B.M. Lal, in his order, said three containers having five kg of MIC gas each be kept under the custody of the district and sessions judge, Bhopal and operative part of the tank pipe may be kept intact till further orders. The government of Madhya Pradesh be informed of this immediately by a wireless message, be ordered in a ruling on a petition filed by a Bhopal citizen, Mr. V.S.S. Murthy, who sought to restrain the government from carrying out the "Operation Faith".

Mr. Justice Lal, who heard the petition at his residence, also issued notices to the state government and Union Carbide (India) Ltd., Bhopal to show cause as to why this petition be not admitted for hearing on payment of process-fee within three days, for a date to be fixed by the office. This notice be made returnable within 15 days, he observed.

Earlier, the divisonal commissioner, Jabalpur, informed the court that the operation had already begun and, therefore, the stay part of the petition had become "infructuous".

Mr. Justice Lal, however, said the petitioner, if he was so keen, ought to have approached the court at an earlier opportunity, but it appeared that he was "sleeping over nights, if any" and had not approached this court "well in time". By now, he observed, the government had made all possible arrangements and incurred heavy expenditure there. If "Operation Faith" was stayed, that would amount to waste of public money, he said.

The government, the judge said, had already decided on neutralising the gas with the consent of technicians and scientists and taken possible precautions and also informed the residents of that vicinity to vacate the area. Thus, the government had shown its bonafides for making the necessary arrangements for "Operation Faith", he said.

The M.P. high court modified the order dated December 16, 1984 passed by the vacation judge Mr. Justice V.M. Lal on a writ petition filed by a Bhopal citizen, Mr. V.S. Murthy, seeking to restrain the state government from carrying out the "Operation Faith" while admitting their petition for hearing.

The modified order was passed on two applications moved by the deputy advocate-general, Mr. M.V. Tamaskar, and counsel appearing for the respondents, the state government and the Union Carbide India Ltd., before the vacation judge. In their applications both respondents pleaded that it was not possible to collect samples of five kg each as they did not posses containers suited for storing such a quantity. The state government, however, stated that, for purpose of analysis, samples of 300 grams of gas could be kept in containers.

Mr. Justice Lal, after hearing the arguments and expert opinion on the subject, modified his original order and added that during the "Operation Faith" atleast three samples of MIC gas be taken under the supervision of an expert committee. The report of analysis should be kept in sealed cover under the custody of the district and sessions judge of Bhopal.

According to the order, MIC gas is to be preserved in five containers of 300 grams each. The containers are to be kept in a special room of the company equipped with water sprinklers. They should be in the custody of the district and sessions judge of Bhopal. The court ordered that the "Union Carbide factory shall not function in any manner whatever until further orders."

No alteration or addition shall be made in all three tanks which were constructed for storage of MIC gas until further orders.

"Its defective safety valve shall be kept in tact as it is until further orders". Mr. Justice Lal also ordered that no other changes should be effected in the apparatus of the concern. He listed the case for January 2, 1985 for final disposal.

The Madhya Pradesh high court on 21 December, 1984 rejected the applications moved by the MP government and the Union Carbide of India Limited, Bhopal, the respondents seeking further modifications of orders passed by the vacation judge, Mr. Justice B.M. Lal on December 19, 1984 relating to the "Operation Faith." The modified orders were passed on a writ petition filed earlier by a Bhopal citizen, Mr. V.V.S. Murthy, which sought a stay of the operation.

The petition was admitted by the high court for hearing and had been listed for final disposal on January 2, 1985. Justice Lal, while dismissing these modification applications, directed the respondents, to strictly follow and observe the right conditions mentioned in the modified order pronounced by him on December 19, 1984.

Mr. Justice Lal further observed that his modified order was passed as agreed and desired by the respondents and after perusal of the opinion of Dr. Shivraman, an expert.

For a last minute evacuation in the event of a contingency, two special trains had stood in readiness at the railway station with their engine steaming and the crew in their appointed places. Similar arrangements had been made at two nearby halts, Diwanganj and Mandideep. Columns of army one-tonne trucks equipped with gas masks patrolled the city for the same purpose as the deadline at the US multinational plant drew closer. Yet all was set for the start of "Operation Faith," the message gave the genesis for the code-name. It said:

> "People's faith has been rudely shaken by the events of the last few days. That faith has to be restored. It shall be restored and none but God himself will see to it, Dr. Varadarajan announced. He further added that there was nothing wrong in having faith in God."

"Is it a wrong name?" Countered questioned the chief minister when cornered by newsmen at the plant gate the next day. "But do such beliefs have any place in the present day world of science and technology," asked a cheecky reporter.

The chief minister answered:

> "I believe that science has a great role to play. But there are limits to science. And beyond these limits, it is only

faith which plays." The operation faith took off to a smooth start half an hour before schedule—at 8.00 a.m. on December 16, 1984.

Tension-bound for the last long two weeks of early December 1984, Bhopal finally heaved a sigh of relief as "Operation Faith" got off to a smooth start and the process of conversion of the remaining quantity of the dreaded MIC in storage-tank of the US multinational plant made smooth progress. Everything was under control and nothing went wrong.

All was well inside the plant and the fear of another disaster receded from every heart in Bhopal. Conversion of the MIC gas into pesticide progressed smoothly and the operation continued beyond the stipulated hours and days. More than half of the toxic gas was neutralised at the end of the day. At the rate, the 15 tonnes of MIC was used up in less than the targeted five days. The idea was to finish off the lethal chemical as quickly as possible and to tell the people that the danger was really over so that those who had left Bhopal could return quickly. Citizens of Bhopal congratulated Dr. Varadarajan and other scientists for successfully converting more than half the deadly MIC into pesticides in "Operation Faith." Meanwhile, the chief minister told the newsmen on December 17, 1984 under no circumstances would the US multinational plant be allowed to remain on the soil of Madhya Pradesh. Finally, the plant who numbered around 632 were being reinstated in other public sector organisations. The Bureau of Public Enterprises and other concerned agencies at the centre had directed various public sector units in the state to consider employing displaced Union Carbide employees on priority basis.

Meanwhile, the government suspended the manufacture of MIC based pesticides as a sequel to the Bhopal gas tragedy. The decision followed the recommendations of Dr. Varadarajan Committees which enquired into the gas leakage at the US multinational's plant. There were 17 MIC-based pesticide plants in the country. All of them were in the small-scale sector and functioned on the MIC supplied by the US multinational from its Bhopal plant. The Varadarajan committee also urged a review of pesticides making processes employed by small scale manufacturing units. The full findings of the committee were still under government examination and had not yet been made public. The committee was understood to have suggested considering new generation of pesticide-making procedures employed in advanced countries for adoption in Indian units.

The industries minister of Madhya Pradesh government informed the House that 51 industrial units and other undertakings in the state were found to be using hazardous chemicals, according to a survey. They included the Bharat Heavy Electricals Limited in Bhopal, the Bhilai Steel Plant, the Bharat Aluminium Company in Korba, the Government Opium and Alkaloid Works in Neemuch and Water Filteration Plants in Indore, Khargone, and Dhar. In the private sector, several chemical plants and textile units, notably in Indore, Dewas, Nagda, and Ujjain, handled hazardous chemicals. Indore alone accounted for 17 such units. The number of units in Nagda was seven and in Dewas six.

The Minister admitted in his written reply that these units had not informed the authorities about the use of hazardous chemicals in their plants. The state government had set up a committee of officials of the labour and agriculture departments and also of the state pollution contral board to conduct a statewide survey to identify units using hazardous chemicals.

The state government had sought the help of Engineers India Limited in New Delhi in its efforts to inquire into equipment and the operational safety of some major industrial units in the state. Efforts were also being made to ensure that effective safety measures were adopted by such units. The factory Act was being reviewed for amendment of rules and procedures regarding industrial safety regulations.

On positive fall-out of the horrendous Bhopal tragedy is the growing awareness among residents in major metropolitan cities like Bombay and Delhi of the need to organise themselves in order to put pressure on the authorities to enforce proper anti-pollution measures. People of Chembur, a Bombay suburb, have done just that and in doing so have set an important example for others throughout the country. At a public meeting the other day, they rightly criticised in no uncertain terms the Maharashtra pollution control board whose chairman, Mr. C.D. Oommachen, was present, for failing to take adequate meaurses to protect the Chembur-Trombay are shocking, to say the least, that a major public sector concern like Rashtriya Chemicals and Fertilisers (RCF) which has already been warned before, and indeed ordered briefly to shut down its urea plant not so long ago, should once again be lax in its operation. This time there was a leakage of sulphuric acid from one of its plants which, when mixed with the water from hosing pipes, created thick fumes causing substantial distress to those living nearby. Nine RCF employees received acid burns, three of whom suffered serious injury. The leak apparently spread to surrounding areas. The plant has been rightly closed down and in inquiry ordered by the Bombay municipal corporation. Preliminary investigation strongly suggests that poor maintenance may be the cause of the leak.

Ironically, RCF has a better record with respect to safety control and anti-pollution measures than most other pollution-prone units in Bombay. One shudders to think what might be happening elsewhere when safety precautions in the better equipped units leave so much to be desired. Equally disturbing is the news of the chlorine leaks from Union Carbide India Ltd. (Bhopal). The management of UCIL has expressed its regrets and stated that the leaks were "minor as if this ends the safety is too serious a matter to be left to management even in advanced countries are much greater than generally realised, as a recent U.S. survey by a Congress sub-committee has underscored. For another, management tends to have vested interest in covering up accidents that may occur. Monitoring agencies at the state and Central levels have to be given the necessary status, independence and authority effectively to check violations of existing laws and to met out severe punishment to erring firms, is necessary. But it is even more important that citizens' groups should also step up their vigilance.

The chief minister said, "I may be guilty of hundred and one things, but cowardice in the face of duty was not among them." The chief minister stated that "if the allegation was proved in the inquiry, I will retire from public life."

In replying to another question the chief minister said: "I was affected by the gas, but I don't have to publicise this."

Regarding earlier incidents of gas leakages, the chief minister noted that reports regarding every incident that occured in the past were being scrutinised by him. He said he was not aware of the precise date on which the statutory inspection of the factory was conducted during 1984. Asked about the statutory requirements regarding inspection and safety measures pertaining to the US multinational Union Carbide plant, the chief minister said, "I can't tell you off-hand what they are."

Replying to another question by newsmen suggestion of attempts on the part of the government to "cover up things, the chief minister reacted sharply by saying, "hush up at any level is absolutely out of the question."

According to PTI & UNI reports, the state government registered a case against the management of the US multinational for causing death of a large number of persons through negligence which resulted in the leakage of the lethal gas from one of its tanks. The case had been registered under section 304-A of the Indian Penal Code. A first information report was lodged at the Hanumanganj police station specially mentioning the operating staff of the plant on duty on December 3, 1984 when the leakage of poisonous gas occured from one of the tanks.

But though India's worst industrial tragedy may fade from public memory in a few years, experts said the aftermath of the environmental disaster would be felt for decades to come. A senior doctor at Hamidia Hospital, Bhopal's largest medical establishment, said that at least eight cases of still births had been reported immediately following the gas leak till the third week of this tragedy. The birth of six deformed babies had been reported in hospitals in the wake of the December gas holocaust. Five of these births took place at Sultana Zanana hospital according to the health minister. In a written reply to a Vidhan Sabha question, the minister stated that three of the deformed babies delivered by gas-affected women had died in Hospital. This included a death reported from a Bilashpur hospital.

The hospital authorities conducted postmortem examination in respect of 27 still births reported in the last six months of the tragedy. They revealed the possibility of the lethal gas MIC having caused the deaths.

Of the 1,350 babies examined under a programme sponsored by the Indian Council for Medical Research, 15 children were found to be handicapped in one form or another. The condition of 3,000 pregnant women due for delivery in July or August, 1985 were being monitored under the ICMR programme. There was no outbreak of any bacterial or viral disease in the wake of the gas disaster. However, 1,759 cases of jaundice were reported from December till June, 1985. The highest number of cases (322) was reported in February, 1985.

Even the doctors had to cope with the pressure of the gas victims admitted to hospitals; medical experts were engaged in studying the impact of exposures to the killer gas on liver, kidney, brain functioning and blood manufacturing organ of gas victims admitted to hospitals. Prevalent uncertainty on the possibility after-effects had given rise to widespread scare among those affected by the gas holocaust. Sources at Hamidia hospital said among those admitted to the hospital were many "scare" victims. These included doctors themselves. At least three of them, after learning about possible neurological disorders following exposure to the gas, had to be admitted to hospital with complaints about weakness in the limbs. Fifty were later discharged after having been found to have been hysterical.

Another as-yet-unsubstantiated report that gained currency was that among the delayed effects of exposure to MIC gas is cerebral damages. Dr. N. P. Mishra, professor of medicine, denied that among those admitted to Hamidia Hospital in the couple of days since the tragedy were many children with inexplicable cerebral complications. He noted that a number of those who died within three or five hours of exposure to the gas had been found to have developed cerebral odema. This was the result of oxygen deficiency.

Dr. Mishra, however, said many gas victims at the hospital were closely watched for signs of neurological disorders. He added that two or three children who died in hospital in the last 24 hours since tragedy had suffered from, among other things, jaundice, whether this was associated with the gas pollution was ascertained by the medical examination.

Pregnant women brought to the hospital were in "terrible pain" and had to be aborted. "Children in the womb had stopped kicking and bodies were rejecting foetuses". He said traces of cyanate and phosgene had been found in the bodies that had been subject to postmortem examination. An army medical officer added that the gas that had claimed so many lives could be phosgene as MIC just could not be so deadly.

Asked to comment on the army officer's version, the Hamidia doctor said MIC gas tended to turn into phosgene when exposed to air due to natural detoxification. The doctor said survivors had inhaled the gas "in more than minimum quantity" would definitely suffer from lung and eye problems in the coming years.

"We really can't say because there has been no precedent of this kind in our country", he said. "But there would surely be after-effects. People, still flocking hospitals and medical camps set up in various parts of Bhopal, complained of nausea, giddiness, itching and burning sensation in the eyes and chest pain. Many victims came frothing in their mouths.

Meanwhile, at the cremation grounds of Chhola Road and other places, the local authorities were engaged in the disposal of heaps of ash accumulated in the wake of mass cremation in the last one week since the tragedy. Gunny bags filled with ash were transported in trucks from Chhola Road to be consecrated to Narmada river.

In Old Bhopal city, though many shops continued to be closed for the fourth day of "Operation Faith", life was fast returning to normal in other parts of the city. All celebrations in connection with the birth of the Prophet Mohammed to mark the Id-ul-Milad-ul-Nabi had been cancelled in view of the ghostly happenings in the city. Government officials were disbursing cash doles to the victims of the tragedy at various hospitals. Almost all slum dwellers close to the US multinational Union Carbide plant silently marched from their colonies to the residence of the chief minister to protest against the inadequate relief provided to them. The slum dwellers, led by 65 year old municipal corporation counsellor, Mr. Sumir Khan, included a large number of women. Mr. Khan said more than 4000 residents of one colony had been affected by the leakage of the killer gas. He alleged that hardly any relief had been provided to them and some of them had gone without food for a week. Later, they held a demonstration near the chief minister's residence. The governor, Prof. K.M. Chandy, cancelled all his public engagements during the months of December, 1984 due to the gas tragedy.

In an instant response to the disasterous leak of gas at its Bhopal plant, the US multinational Union Carbide, closed identical plant in West Virginia (USA), where it had been using gas manufacturing process similar to that of Bhopal.

A spokesman of the US multinational said:

> "We do not know what went wrong but it was not as if there were anything unusual about the plant. The poisonous gas, isocyanate, is also made by other companies in many countries".

The US multinational Union Carbide experts suspected that the vent scrubbers that had been installed at the Bhopal plant to prevent leakage of the poisonous gas might not have worked for some technical reasons. The vent scrubbers are a device that would scrub the leaking gas and render it harmless.

In a statement issued from its headquarters in Danbury, Connecticut, characterised the Bhopal disaster as unprecedented. It said the company had been producing the same gas under similar process at its West Virginia plant for the last 25 years without any mishap. The company said it was in consultation with the government of India to find out what relief it could offer to the families of those killed in Bhopal.

The company feared that it would have to face a spate of galclaims filed by the families of the gas leak victims. Legal experts believed that the political liability of the company could be hung but "the company is big enough. It could handle whatever is assessed in India".

Health and occupational rules in the United States prescribe that workers must not be exposed to more than 0.02 parts of isocyanate per million parts of air. This works out at 0.05 mili-grames per cubmic metre. It is not clear whether safety standards are in force in India and if these were enforced at the factory in Bhopal. If they were, the gas would not be stored in quantities larger than a thousand fallons.

The US multinational Union Carbide still held as on April 2, 1985, among their chemicals, 27 tonnes of chlorosulphonic acid and 21 tonnes of chlorine. This was stated by the minister for industries, Mr. Bharat Singh in the Vidhan Sabha. The minister chose to give details of the chemicals stocks held by the plant in response to an opposition plea that lack of information such as the nature and quantity of various chemicals being held by the plant had contributed to the prevailing tension among the people apprehending further gas leakage.

The issue was raised through a notice of adjournment motion. The speaker, Mr. R.P. Shukla, disallowed the motion, brought forward by six BJP members, after the House had spent more than one hour considering its admissability. The opposition members, in the notice of the motion, referred to two incidents of alleged gas leakage and also to the failure on the part of the authorities to give sodium thiosulphate treatment to all gas victims.

The minister observed in this connection, by way of a clarification that the health authorities had not received any guideline from the Indian Council of Medical Research (ICMR) stating that all gas victims should receive sodium thiosulphate injections. They were given it, only if found necessary, after proper medical examination. About 300 persons had been given the injections so far.

Some BJP members wanted the government to make public the list of such persons, alleging that they were all VIPs. The "indifference" of the authorities to the plight of gas victims had caused widespread resentment and the gas Kand Sangarsh

Morcha has threatened to stage a demonstration in protest against the indifferent attitude of the government. Referring to the recent incidents of alleged gas leakage, the opposition members said the fact that such incidents were reported even after the December 2/3 disaster that took a heavy toll reflected "negligence" of the government in matters concerning the Union Carbide episode.

Refuting the opposition charges, the industries minister stated that the government was alive to the situation caused by the stocks of chemicals held by the Union Carbide unit. He said the factory authorities had been directed to dispose of in a safe manner all chemicals in their possession. Bulk of the stocks had been removed from the factory and some of the chemicals, such as monomethyl amine, had been disposed of. Mr. Bharat Singh expressed the hope that stocks of the remaining chemicals would be cleared soon.

Giving the latest stocks position, the minister said the factory was left with only 27 tonnes of the initial stock of 155 tonnes of chlorosulphonic acid. The chlorine stock had been reduced from 110 tonnes to 21 tonnes. As for the recent incidents involving these chemicals, Mr. Bharat Singh noted that the police visited the factory on March 28 soon after receiving a report about a railway employee on duty at a cabin close to the factory having suffered eye irritation. However, no information regarding any gas leakage was obtained.

According to a report from the factory authorities, chlorine smell was felt by those filling a tankar for transporting chlorine acquired by Hindustan Insecticides Ltd. Work on filling the tanker was suspended and the workers on the job, on medical examination were not found to have suffered any adverse effect. The incident did not cause chlorine pollution. The minister noted in this connection that the district authorities had received on earlier occasions as well reports regarding gas leakage. However, such reports were found baseless on inquiry. Mr. Bharat Singh sought the cooperation of members in the efforts to ensure that such rumours sustaining a sense of fear among the people did not gain strength.

The state government, on its part, had taken all possible steps to provide safety, medical aid and other relief to the people in the gas affected areas. UNI adds:

> The chief minister, Mr. Vora, denied that a bribe of Rs. 10 lakhs had been taken by someone at the time of renewal of the licence of the Carbide pesticide plant. The Union Carbide had no hold over the government, he added. Mr. Vora said he would hold discussions with opposition leaders and take them into confidence regarding the December 2 gas tragedy.

Acccording to experts, the one reason why the cloud of leaked gas proved to be so lethal and why it did not dissolve so long could be that it was formed during the night. In the absence of the sun, the forces that would have helped make the cloud of gas dilute and less harmful were not fully in operation. Moreover, the experts believed that the victims must have been sleeping while the gas started having its deadly effect on them and they must have been taken unware of.

Quoting specialists from the national institutes for occupational safety, the New York Times said that 50 per cent of rats that had been exposed in an experiment to the poisonous gas isocyanate died as a result. "Even light exposures have long lasting effect on human beings, causing an allergic or asthmatic reaction to the slightest further exposure."

"No skin exposure is permitted to the gas in American plants. If it is impossible to keep the air level below 0.02 per million, the workers must be protected individually." The question is were there or similarly rigid standards of occupational safety observed at the US multinational's plant at Bhopal.

The Prime Minister Rajiv Gandhi said on November, 7 1985 that industries were taking the responsibility and burden of protecting the environment and he asked them to adopt methods to avoid Bhopal-type tragedies. Industries did not pay adequate attention to the dangers of chemicals and the priorities on pollution should be an integral part of all new technologies that were being developed. Rajiv Gandhi said, while inaugurating the three-day Second World Congress on Engineering and Environment[1]:

> "A little foresight in planning proper safety measures will prevent such accidents like Bhopal and will prevent such high costs. . . If we do not pay a price today we will invariably pay a much higher price tomorrow. We must build this in our system."

Adequate anti-pollution measures must be built at the very first stage of setting up of an industry and particular toxins must be adequately indentified and removed before the plants get going. We have had very bad experience in Bhopal with an air pollution—with a chemical plant getting out of control. Again, it is not just a question of safety measures but a much deeper question of also taking adequate protection measures in case something goes wrong.

People will accept the extra cost of environment protection when it is spread out over all the consumers and also over the full time span of a particular process. When that cost comes at a particular time, as it did in Bhopal, it became a much harder problem, not only in terms of relief measure for the governments that are involved but also in terms of compensation for the industry that is involved.

A shift to low pollution industries that use less natural resources that would damage the environment less should be emphasised. It had been found in India that protecting specific area one was able to protect the environment. By protecting a particular animal like the tiger, we have protected not just the tiger but the whole environment in that area.

The government had meanwhile stopped the transport of highly toxic substance in drums and tankers across the country, but regretted that this still happened in the developed countries stating that no project could be started in India unless it had full environmental clearance. Rajiv Gandhi said that if a particular forest had to be cut to accommodate a project, an

1. Hindustan Times, November 8, 1985.

equivalent area is replaced at some other location preferably touching another forest, as India's forest cover was already too low. One has to balance the cost of environmental protection with development.

Rajiv Gandhi assured that the government would bring about an awareness of protecting the environment and of the dangers of damaging it right from the school curricula:

> "No amount of legislation will bring about type of protection that is required in India. . . for ultimately it has to be the people's awareness that will help protect the environment." Rajiv Gandhi said.

Prime Minister Rajiv Gandhi's stress on a harmonious relationship between man and nature and his exhortation that human values are very important in the developmental process have special significance in the context of certain glaring distortions in the present day life. To some extent, environment has got a low priority in planning. But the factor of pollution can not be ignored for long particularly after the lessons learnt in Bhopal.

The industry may chase profits but it has to pay adequate attention to environmental needs. The damage being done at present by high toxic elements will have to be avoided in any process of planning. This means natural resources will have to be conserved and looked after well. The new social forestry projects may help tackle the issues to a considerable extent. It is also true that progress would not be worthwhile if human values were put on a discount. Modern technology is necessary but the accent should be on the well-being for human development. At the same, our culture, tradition and way of life should not be disturbed. Machines are for human beings and they should not overpower them.

3

RELIEF AND REHABILITATION

THE STATE government had a moral responsibility towards the gas victims who gave birth to incapacitated or deformed children in particular. It is duty of the government to protect the sanctity of human life of its citizens under all circumstances. Immediate steps should be taken in Bhopal to monitor the births of all infants and make an urgent assessment from hospitals, nursing homes and affected areas to gauge the enormity of the problem and start discussions on the problems of compensations. If the Bhopal disaster had occured in some other country in the West where the right to withdraw treatment rests legally with the doctors as in the United States then perhaps the demand for euthanasia would have been more vehement. In India, the traumaog giving life to sustaining the lives of these children is something that only the gas victims know themselves.

It will be decades before the genetic effects of MIC exposure are known, after children who had been scientifically declared deformed by the gas grow up and give birth to the third generation MIC victims. Then the perpetuating physical agonies of these children and the worst fears of doom would come true. It was feared that more birth-defective children were born by the end of the third timester at the end when gory picture of such congenital malformations were complete. The long term mental and physical effects of MIC exposure on the gas victims had set a unique precedent since it was only the people of Bhopal who became the unwitting preys to the unknown killer. The prospect may not be so macabre and the birth of "thalidomide children" in the early Seventies, but these malformed physical weaklings were a reminder of the thalidomide case which stunned the world years ago into realising the effects of unknown chemicals and drugs on new-borns. The children of the victims of thalidomide, a sleep-inducing drug administered to thousands of pregnant women in Europe were grossly deformed. Their hands and feet were worst-affected. Some of the children born to the victims of Bhopal had severe congenital defects, including retardation of organs like the lungs and heart and the softening of the brain tissue—children who would live but remained only marginally alive.

The Bhopal tragedy raised two important questions about how much compensation should be provided to the 2,50,000 affected people and what compensation should be provided to the victims and those who were injured or maimed. And finally, what deterrent should be devised to Union Carbide for its callous negligence.

Soon after the tragedy occured the then chief minister Arjun Singh had said that the answer to the vexed problem of sorting out the injured was a door-to-door survey conducted by the Tata Institute of Social Sciences (TISS), Bombay, assisted by other social work institutes. But it did not detail out that manner. In the first place, TISS surveyed just over 25,000 families and was to return a few weeks later for the rest but TISS team leaders were frustrated by the indifference of the state government towards the exercise. The work was yet to be completed in the area field survey. Moreover, the state government had still to complete the processing of the data so collected by the Census Department as well.

The outcome of the exercise was very significant in a sense that the interim cash relief could be extended to the victims only after the processing data was recorded. Families of the dead were to receive Rs. 10,000 for every member lost but of 1,762 certified as dead so far—the figure would rise to 3000 mark finally—only 848 of them so far received the money. According to government sources, many of the remaining claims were still unpaid either because the heirs were not available or because there was difficulty in the identification of the dead. Initially, the government had announced ex-gratia payment of Rs. 2,000 to the seriously affected and Rs. 100 to Rs. 1,000 to those slightly affected. But the government, meanwhile, changed its policy and fixed payment at Rs. 1,500 per affected family and not per person—which had a monthly income less than Rs. 500.

Though about 18,000 families were expected to ultimately fall under the category prescribed only about 9, 00 families had so far been identified and of these only 400 families had actually received ex-gratia payments. The state government had yet to answer to important questions:

1. What is the affected area? and,
2. Who is an affected person?

Once this parameter was determined by the government, the extent of claims for compensation payment should be known. Meanwhile, a Directorate of Claims had now been set up to adequately record claims of the gas victims as well as the claims of commercial establishments, shops and industries. It was stupendous task for the Directorate to document all the claims from at least 2,50,000 people and organisations.

According Dr. Ishwar Das, additional chief secretary and relief Commissioner, "some 800 small-scale manufacturing units and some 18,000 to 20,000 shops and establishments had seen their business hit in the aftermath. "From another official sources, it was learnt that the government chose to utilise the existing "step-up" scheme for self-employment and under this scheme, about 4,000 cases had been forwarded to the banks for loans but few people so far had received the money.

The affected people had been considerably benefited with the free supply of foodgrains worth more than Rs. 10 crores. Four hundred and seventy tonnes of sugar, 4,00,000 litres of edible oil and 3,90,000 litres of milk. Wheat (27,000) tonnes and rice (8,100 tonnes) were continuously supplied to the affected people. With the closure of the plant, about 850 employees were rendered job-less. Neither the government nor the management of the plant had seriously so far considered the fa e of the unemployed workers. Attempts were being made by the government to accommodate Carbide workers in the public sector undertakings. But the workers expected to be absorbed in Bhopal itself. The trade union challenged in the High court the manner of the closure of the plant:

"Carbide won't be allowed to operate in Madhya Pradesh ever."

The governor, Prof K.M. Chandy, accused the Union Carbide of having "shown callous unconcern" for the gas affected. He noted that the rehabilitation of gas victims was primarily the responsibility of the multinational corporation. The governor, who laid the foundation stone of a 100-bed hospital, to be set up to provide long-term medical relief of MIC patients, said the state government would leave no stone unturned in its rehabilitation efforts. The decisions to construct the hospital at an estimated cost of Rs. 2 crores was announced by the chief minister, Mr Motilal Vora, earlier 4 months, when he visited the 30-bed hospital set up at a residential building near the Union Carbide unit.

Operation under way had to be seen to be believed. And not even a vague plan had been prepared so far for the rehabilitation of the gas victims—the tens of thousands who lost the wage-earners of their families, and had been widowed, orphaned, crippled maimed or rendered too ill to be able to work. There is no private relief initiative worth the name in Bhopal. The numerous organisations which offered some assistance to the victims of the Union Carbide gas leak in the first days of the catastrophe have for the most part wound up and left the city. They treated the affected people as if they were victims of a natural calamity such as flood or an earthquake, with no appreciation of the crucial fact that MIC was a deadly long-acting chemical poison whose effects were as lethal as they were chronic, persistent and in some respects irreversible.

The Central and state governments' approach had hardly been different. The Centre had not treated the catastrophe as a national tragedy and left the task of providing relief and rehabilitation to the Madhya Pradesh government. The latter, for its part, had chosen to adopt a minimalist, ad hoc, miserably unimaginative approach to the whole issue of relief. It had been content to dole out small sums of money, such as Rs. 500 or Rs. 1,500 (Rs. 10,000 in case of death in a family) to those who had been seriously affected by exposure to MIC. It had instituted a system of providing free rations to people. But there was reason to believe that over 2.5 lakh affected cases and over 1.5 lakh people who were in urgent need of help had not received any kind of government relief. For the most part, the relief package had consisted of the free supply of 12 kg of cereals, $\frac{1}{2}$ kg of oil, $\frac{1}{2}$ kg of sugar per adult per month and 200 ml of milk per child per day. The state government said it had spent a total of Rs. 38 crores by way of relief and compensation so far as on October 1985.

While the quantum of relief would appear substantial at first sight, it was far from adequate in reality. First, the supply was erratic, and not enough to meet the special nutritional requirements of those who were ill. Secondly, at least a quarter the of the quantity recorder as being doled out was not reaching people. In some of the badly affected *bastis*, the proportion was as high as a third. Thirdly, the quality of a food items suppiied was extremely bad—in some cases, as much as a third or four-fifths of the wheat was stones, pebbles and other muck, and the milk was liberally adulterated with water. And finally, several thousand poor families who desperately needed support did not receive any free rations because they had not been issued ration cards (as they live on un-authorised plots of land) or had been tricked by local ration-shop owners and petty officials into surrendering them. They had been turned into beggars.

FANCY PLANS TO AID GAS VICTIMS

The National Action Committee of the Bhopal Gas disaster described the total Rs. 322 crore plan of Madhya Pradesh government for gas victims as a cruel joke on the poor sufferers. The chairman of the committee, Dr. Sujit Das said that the plan envisaged a beautification of Bhopal and had little to do with the welfare of gas victims:

"off the total Rs. 322 crores, a mere Rs. 38 crores had been allotted for the welfare of gas victims. The rest was to be spent on fancy schemes like construction of a new airport, railway and bus stations, setting up of a botanical garden, etc. The government was spending a mere two paise in a rupee on community development in the gas affected areas in the plan: "We have nothing against the government constructing a new airport but do not do it in the name of gas victims," he said.

The Madhya Pradesh chief minister, Motilal Vohra directed the state administration to drop fancy schemes like development of airport, botanical garden and swimming pool from the state's draft plan to provide funds for the gas victims. His directive followed the hue and cry raised by various sections of society which strongly criticised inclusion of such schemes in the plan at the expense of welfare of the victims. He gave the instructions at a meeting of senior officers. It was also decided to set up an institute of disaster mangement in Bhopal in memory of the gas victims. Great emphasis was being laid on the economic and social rehabilitation of the victims. For this purpose work sheds-cum-training centres were set up to provide the gainful employment. At the same time full support was given for providing raw material as well as to arrange sale of the products. It was also proposed to set homes for distressed children on the lines of ashram schools in tribal areas. Such children whose parents were not able to look after them because of failing health would be provided board and lodging in the hostels attached to educational institutions. It was proposed to set up six such ashrams. In addition 900 creches and anganwadis were also planned to be set up. Setting up of 30 community centres in the affected wards was also part of the plan. The gas victims, who were no longer capable of doing hard physical labour, would be helped in acquiring new skills. For imparting skills of various descriptions to the gas affected people or members of their families, vocational training institutes, three industrial training institutes for boys and two for girls, and training centres would be set up. According to M.N. Buch, the major break-up of the proposed expenditure was Rs. 41.97 crores for creation of an urban infrastructure; Rs. 52.68 crores for development of community facilities; Rs. 92.83 crores for housing; Rs. 95.86 crores for conservation and environmental development; and Rs. 38.85 crores for urban re-development and rehabilitation of gas victims.

The plan included the construction of a new airport at Bhopal, redevelopment of the railway station, construction of 11 college buildings, introduction of an eletric trolley, bus and tramway system, construction of swimming pool and parks, development of areas of natural beauty and conservation of archaeological and historical places and so on. Purely as an after-thought a sum of Rs. 1.72 crores would also be spent on development of vocational and industrial centres.

Those who died in the worst industrial disaster of December, 1984 were beyond help, but the tragedy also left behind two and half lakh people whose health was affected. One guess was that about 60,000 people representing 15,000 families and, therefore, 15,000 bread winners suffered such physiological and psychological damages that they were handicapped in putting in day's work. Their earning capacity was badly eroded. These were broken people, people who had not only lost members of their family but were also faced with the grim prospect of never being able to rehabilitate themselves in their former occupations. They would be reduced to beggars by keeping them on a dole. Their innate human dignity demanded that they be helped to become useful members of society once again. Instead, an unthinking and cynical government produced a plan for city beautification and development, with hardly any component at all of rehabilitation; in economic terms the poor, miserable, mangled remained of innocent people whose lives had been blighted.

It must need a bureaucratic mind of supreme insensitiveness, trained by years of file pushing to produce such a plan. But what was more surpising, the political masters of these bureaucrats had also not seen through the game and in fact were proclaiming from the house tops that they were doing something wonderful for the gas victims. This was something of which the Prime Minister must take note of. The plan prepared by the Madhya Pradesh government deserved rejection outright. The majority of the gas victims were unskilled, semi-skilled and at the lowest rung of the economic ladder. They now found themselves unable to perform the manual tasks through which they earned their living. What was needed was a very detailed survey of each wage earner to try and ascertain his existing level of skills, the degree of physical impairement, his capacity and inclination to learn new skills and the capability of training organisations to impart such skills. It was on this basis that a training infrastructure would have to be evolved and plans drawn up either to mitigate the physical handicap in the present occupation by making available mechanical aids, or to create a new job infrastructure which would shift people from manual work to relatively easy bench assembly jobs.

Obviously, the new employment infrastructure would have to be designed to suit the skill and aptitude levels of the affected persons so that it was they and not outsiders who derived benefits. The exercises involved in this task were extremely arduous because they would require a man-to-man and case-to-case study. They had never been attempted on such a large scale anywhere in the country and that was why most of the voluntary organisations and institutes of social studies avoided them in Bhopal. It was not impossible to find that funds to create a new employment infrastructure in the city. Certainly the task would be less costly than the Rs. 322 crores proposed to be spent on utterly harebrained schemes. It would be possible to involve entrepreneurs also in creating new job opportunities, provided we could tell them what was expected of them.

Inadequate as it was, even this threat of suspension had always hung over the free rations. Each month, the government extended the scheme on a purely ad hoc basis by another month. The last such extension was decided upon prior to the elections to the state assembly. Now that these were over and a new government had been installed, no one expected the relief scheme to continue unless pressure was brought to bear upon the state administration by the Centre and by the affected people.

Meanwhile, all hopes that a well-planned and systematic relief and rehabilitation programme would be launched remained shattered. These hopes were raised when the state government commissioned the Tata Institute of Social Sciences of Bombay to conduct, in collaboration with other colleges of social work, a general survey of the affected population. This survey was suddenly suspended and a number of localities exposed to MIC were excluded. Since then the work had been resumed on an ad hoc basis by the state government but the total number of people covered was less than 1.5 lakh, much less than the over 2.5 lakh persons affected.

The TISS study came in for sharp criticism from social workers among others, on the grounds that it was poorly designed, that in questionnaires were filled in a haphazard and unreliable manner, that a number of deaths and serious illness (amongst those people whose houses were found locked or deserted nearly 300 dwellings) were excluded from coverage, and that untrained social work students were quite irritationally asked to fill in an ICMR questionnaire on the number and nature of serious illnesses—something that non-medical person was certainly not qualified to do. However, Mr. Arjun Singh, former chief minister, had announced that the survey would from the basis for determining compensation claims of the victims.

The decision may be changed but the basic ad hocism of the government's approach persisted. At any rate, the TISS survey provided the basis of the state's plans to rehabilitate the widows and orphans, which were the only target groups to be covered although they formed a minuscule proportion—less than one quarter per cent—of those who had been in distress. As things stand, however, the state government had no plans whatever even for 318 widows and orphans. Nor was it under pressure from the opposition parties or voluntary agencies to launch a serious rehabilitation programme.

The mood of the people in the affected *bastis* of Bhopal remained one marked by despair and a profound despondency. A new element, however, was anger at the government's indifferent attitude. There was resentment also against the talk of out of court settlement.

The official relief outfit was poorly equipped and understaffed. Even coordination between different agencies—medical, legal, relief, and the finance and civil supplies departments, etc.—was virtually non-existent. All in all, the Union and state governments failed to cope with the aftermath of a grave industrial disaster involving a chemical poison.

And yet, felt a section of the bureaucracy as well as social workers, it was not too late to set up a high-powered authority to deal with the situation, along the lines of what had been done at Seveso in northern Italy in 1976 following an explosion in a pesticides plant which released a cloud of dioxin, a deadly cancer-causing poison, into the air.

Within days, an "uffizio speciate", or a powerful special bureau was set up at Seveso. It comprised top-level personnel from various different fields, government agencies, medical and toxicological experts. The bureau functioned as an apex co-ordination agency, well-funded and authorised to make a whole range of decisions, from evacuating the worst affected zone to treating people, from commissioning toxicological research to decontaminating the area, from providing relief to preparing the basis for law suits against Montedison the firm reponsible for the accident.

It was only because such a massive initiative to cope with the emergency was launched that the damage caused by dioxin at Seveso could be contained. It was still not too late—although it soon would be—to establish a high-powered, Central government backed special authority to deal with all the problems—medical, legal, relief, counselling, physical and psychological, rehabilitation, compensation, etc.—posed by the aftermath of the world's worst industrial disaster. A great deal needs to be done with vigour, with speed, with human concern and with missionary zeal if the destruction wreaked by MIC was to be arrested or contained.

Apart from a handful of skilled operators who had taken up jobs in C.J. Gelatine, the rest of the workers were jobless. In India, that was not often a fate deserving of special sympathy.

But for the Carbide workers, contented inhabitants of a high-wage island, the gas leak and its aftermath had been a continuing trauma. Hatim Zariwalla, president of the Union Carbide Workmen's Union, said with a rueful smile:

> "Right now all we have to look forward to is the day when our bonus will be paid". Added Kishan Prasad, secretary of the union: "Those 45 minutes (when the gas leaked) have changed our lives totally. We have lost our mental peace."

The workers thought they were caught between a heartless management and indifferent government. The Carbide management offered them compensation as per the letter of the law (15 day's salary for every year of service) but they all refused to accept it. Prasad, who pointed out that the workers had been inhaling for the last five years the deadly methyl-isocyanate which killed thousands, explained: "We think the company is not doing its moral duty."

Their complaints against the Government were more varied To start with, they contested the validity of Section 25 (c) of the Industrial Disputes Act under which the government allowed Carbide to close the unit. They had even filed a writ in the high court at Jabalpur on this but they realised that the closure of the unit was a fait accompli; on government would, for they sake of their jobs, earn the people's wrath by reopening the factory. They were more upset with the government for not doing enough to get them jobs. Arjun Singh, they said, did nothing in the three months that he was in power than get them registered in the employment exchange.

Even the current chief minister Motilal Vora took up the problem seriously only after the closure. Now a bureau had been formed, comprising four secretaries to the state government, which was entrusted with finding jobs for the workers. The setting up of the bureau had certainly expedited matters. The bureau had offered jobs to over 200 workers, according to Hrishikesh Mishra, special secretary (labour). The problem was that the workers thought the jobs were not good enough. Carbide used to pay them exceptionally well. After overtime some sweepers were known to gross over Rs. 2,000 a month. In comparison, the Madhya Pradesh Housing Board had offered to take 22 people as Khalasis and peons on a basic pay starting at Rs. 450 a month in Carbide they were designated helpers and paid Rs. 1,200 per month.

Mishra felt, with some justification that the workers had yet to reconcile themselves to the realities of the situation. The workers, he said, were more busy trying to prevent the government from agreeing to the closure. All the same, he maintained every single one of the 612 workers who had registered with the government had been offered atleast one job each. But the workers were too fussy: the J.K. Raymond cement factory, for instance offered jobs in the range of Rs. 1,400-1,600 for 32 people but the workers refused to accept them because it meant going to Bilaspur. "I can only take the horse to the water, I can't force it to drink it", said Mishra.

"The Carbide workers should realise that there were some problems in their invading any other organisation", he added. In the Madhya Pradesh Electricity Board, which had offered jobs to 31 people, for instance, the management was reluctant to take them because there was an agitation brewing among daily workers against taking outsiders for vacancies which they would otherwise have filled.

On its part, the Carbide management was typically hard-headed. Sheo Mitra, the personnel manager at the plant, told a correspondent: "We could not have offered compensation beyond the law. In any case, we were not approached for a negotiated settlement by the workers on this issue". Asked if Carbide did not feel that it was morally bound to go beyond the latter of law, Mitra pointed out that Carbide had suggested several alternatives for providing jobs, like offering to run the closed battery unit of the J.K. group, but the state government, for reasons unknown, had not shown any keeness to associate itself with Carbide.

There was little to relieve the workers' gloom. Government officials said that compensation was not a closed issue but there were no positive singals from the Carbide management. Representatives of the employees met with the management in Bombay on July 25, 1985 but the demand for better compensation fell on barren grounds because of "diverse pressures on resources", said management sources. Union office bearers said they now hoped "to get justice from New Delhi". The workers could not afford to lose hope; what they realised, however, was that their days of relative affluence in Carbide were over for good.

The Central government decided to create two funds to deal with claims and to render relief to the victims of the Bhopal gas disaster. The claims and relief fund would deal with the amounts received in satisfaction of the claims and any other amounts made available as donation for relief purposes. The fund would be used for the settlement of claims or as relief to persons who were affected by the tragedy. It would also be utilised for disposal of claims arising in future or for disbursal of amounts to the Madhya Pradesh Government for the social and economic rehabilitation of the persons affected by the disaster.

A gazette notification had been issued for setting up the Bhopal gas leak disaster (registration and processing of claims) scheme. The Second fund—the processing of claims account fund—was created and maintained by the Central government. It will include the amount which the Centre might, after due appropriation made by Parliament by law, credit to the fund.

The Central government would determine the total amount of compensation to be apportioned for each category of claims and the quantum of compensation payable in relation to each type of injury or loss. In determining the compensation several factors, including the probable life span of the person affected by the disaster, the actual or projected earning capacity of the person affected and the likely expenditure on immediate and anticipated medical treatment would be taken into account.

A major challenge faced by the Madhya Pradesh and Central Governments was how to rehabilitate the helpless victims of the Bhopal gas tragedy. Some 10,000 persons would have to be provided alternative jobs because they were not capable now of doing hard physical labour.

After initial period of make-shift arrangements in the hospitals when petients had to be kept in tents for want of space, the medical relief was properly organised. Sixty beds were reserved in the Hamidia Hospital and most modern equipments were provided to monitor the condition of the patients. Sixteen dispensaries were opened in the seriously affected areas to look after the day-to-day problems to the sick people. One eye hospital in the collaboration with Royal Commonwealth Society for the Blind had also been set up. Eight ambulances were exclusively for use in the affected areas. One hospital with 30-beds had been set up near the seriously affected areas. The work for adding another 70-beds to this hospital, at a cost of Rs. 2.50 crores had already been underway. Medicines and injections were given free of cost to the gas victims. Those who were still suffering from the irritation in the eyes were supplied spectacles without any charge.

The shock of losing their near and dear ones and the horror of the night, as was to be expected, caused acute mental depression in many cases. The National Institute of Mental Health and Neuro Sciences prepared a document on how to care for such patients and trained doctors were looking after mental cases.

Pregnant women were being tested by means of modern equipments including ultrasonograph. Special care was being taken of such women by supplying nutritious food to them through the 343 Integrated Child Development Schemes operating in these areas.

As had been mentioned earlier, the mix of gases which leaked included Hydrogen Cyanide and other toxic cyanides. The autopsy performed on the dead also showed the unmistakable signs of cyanide poisoning. Some experts, therefore, suggested the administration of sodium thiosulphate injections. Sodium thio-sulphate was antidote for cyanide poisoning.

Some experts believed that cyanide poisoning could be chronic and hence sodium thiosulphate injections should be given to the entire affected population. Other experts, however, believed that, cyanide in human body should either kill a person within 48 hours or it should be excreted. It could not last in the body beyond 48 hours. They held that there was now no need to giving sodium thiosulphate injections to the sick people.

The whole controversy was submitted to the Indian Council of Medical Research for its opinion. The issues were cebated in three sessions of the Council in the presence of such eminent doctors as Dr. Ramalingaswami and Dr. Shriramachari Director-General and additional Director-General of the ICMR respectively.

The ICMR then laid down certain guidelines for the use of sodium thiosulphate injections. It was felt that under only those circumstances which were mentioned in the guidelines should this drug be injected. It was thus left to the discretion of the clinician whether the drug was needed or not. This was how it should be.

Unfortunately, a great deal of controversy had been raised about this matter. On this issue the guidelines given by ICMR should be strictly followed and no one should be compelled or persuaded to take sodium thiosulphate injection unless prescribed by the physicians. A great deal to stress was now being laid on proper exercise for regaining the capacity of the lungs affected by MIC.

Thousands of people lost a substantial part of their lung capacity due to the reaction of MIC with air sacs in the lungs. The Indian Council of Medical Research is conducting several studies to determine the long term effects of the gas. Meanwhile, the government was monitoring condition of the patients by providing them a card for permanent medical record.

It was estimated that approximately 10,000 persons would have to be provided alternative jobs because they are not capable now of performing had physical labour. Various schemes for permanent rehabilitation of such persons were being finalised. Some of the severely affected people would be provided shops in a shopping centre specially to be developed for such people.

Certain small-scale industries which could be carried out either at home or in shops to be built in nearby areas, would be taken up. The women were being trained in sewing and zari and bead work. The porters in the railway stations would be given trollies so that they did not have to carry loads on their heads.

Voluntary agencies working in the gas-affected localities reported a number of cases in which the victims were in a state of acute depression. They greeted social workers and other visitors with a blank stare and stony silence. Efforts by social workers to give them drugs for mood elevation or talk them out of depression had limited success. The problem was said to be far too extensive for individual social workers with limited resources to handle. In the absence of institutionalised efforts to cope with the problem, the languishing gas victims, physically incapacitated and mentally drained, so far remained untouched by the official relief programmes. There was no one to fetch them relief ration or take them to a make-shift clinic set up in their neighbourhood. Many of them just did not think of the relief provided by official and voluntary agencies. Some were said to have lost their will to live. According to a conservative estimate, at least 1,000 families in localities such as J.P. Nagar and Kanchi Chhola were too weak and helpless to avail of the relief offered by the government by way of distribution of ration and medical help.

Official sources said steps to cope with trauma effect and behavioural disorders among the gas affected had been discussed at a meeting of medical specialists from various parts of the country held on January 15, 1985. The meeting was part of continued efforts of the government to review line of treatment provided to hospitalised gas victims. Nearly 25 specialists from leading medical institutions in New Delhi, Bombay, Madras, Calcutta, Ahmedabad and Lucknow attended the meeting.

As for the gas victims not covered by existing institutional medical care, extent of the problem was being assessed through a survey of the gas affected areas being conducted. Among the aspects covered by the survey was the health status of the affected population and the morbidity pattern in the gas affected localities. The survey was completed long time ago.

But the gas victims, uncared for and bed-ridden in their houses, could not be expected to wait till the formulation of a long-term health care programme on the basis of the survey report. It had been suggested that special medical squads should be set up to identify patients requiring intensive medical attention in their houses. In the J.P. Nagar locality, for instance, there were such patients in many jhuggis. Because of domestic circumstances they were neither inclined nor in a position to go to hospitals.

The condition of the patients indentified for daily house visits could be monitored and documented for preparing a detailed health history of the gas victims. Apart from provising medical evidence for purpose of compensation cases to filed on behalf of these gas victims, the case history of the patients would facilitate data requirement for proposed studies on long-term effects of exposure to the poisonous gas (MIC). Along with a programme of intensive medical attention to the needy families in the gas-affected localities, it had been suggested, the authorities should make arrangements for provision of free ration and firewood and for registration of the Victims by the legal aid clinics as part of a package programme for door-step relief to the helpless among the victims.

The gas tragedy had horrified the whole world, but, seemingly, not the conscience of the top brass of Union Carbide company and its administration. On the contrary, the political, civil and medical systems were either obliged, purchased, persuaded or pressured by the Carbide managemement to abet the killer company. From the dreadful night of December 2/3 1984, every development clearly pointed for the machinations of Union Carbide to cover its tracks.

Wide publicity was given to the fact that the plant at Bhopal had finally been closed down. This was exactly what Carbide had been planning for the last two years. In 1983 the company's management had literally abandoned the plant, which had ceased to be a money-spinner for them.

But after the virtual murder of more than 15,000 people (though the official figure still stands at 1,978), crippling of several thousands more and leaving an entire generation with horrendous side-effects, the plant finally closed down from July 9, 1985. As a parting kick to its employees, Carbide offered "15 days salary for every year of service." This was worse than rubbing salt into open wounds. The employees rightly rejected this offer. The closure, however, resulted in 630 workers, and their families being reduced to penury—apart from the 200 other workers employed in small ancillaries attached to the plant. The question was, what could be done under the circumstances? Yes, the plant had to be closed down, but the government should have handed it over to the employees to run the non-toxic production of any other items, this beidg run by an employees' cooperative. To add insult to injury, the State Labour Department has said that the Bhopal plant of Union Carbide had been closed down due to a "dispute between the management and its employees."

The confusion had been further confounded by the controversy over the use of intravenous injections of Sodium Thiosulphate. About 12,500 affected persons had been admitted to different hospitals. It went to the credit of the physicians of the Medical College and other hospitals that deaths within their premises were only 446. This proved that the symptomatic treatment given at the time, with the liberal use of life-saving drugs like Corticosteroids went a long way in saving lives.

The controversy actually started when a West German, Dr. Daunderer, air-dashed to Bhopal with hundreds of ampoules of IV Sodium Thiosulphate. There was little literature available about the toxicity of MIC gas, thus physicians, led by Dr. N.P. Mishra, Head of the Medicine Department, Gandhi Medical College, were reluctant to accept this therapy as they were not convinced about the theory of cyanide poisoning.

On the other hand, Dr. Heeresh Chandra, Forensic Department head, insisted that he had found signs of cyanide poisoning during the postmortems. He preferred the administration of IV Sodium Thiosulphate and had, in fact, administered it to many patterns. The ego-clash between the two doctors turned the hospital into a battlefield.

This was not to suggest that all was well within the medical department. From the very beginning, it had been changing its stance. First, it said that the killer gas was phosgene and not MIC; it later reversed its statement. It was now being said that since the temperature within the tank had risen above 4000°C., MIC had split into different side-effects. Further, it had been asserted that the worst damage done to the victims was during the first night only.

This lie had been nailed by the fact that numerous cases were reported daily of abnormal children being born, lung complications, increased debility, loss of vision and even paralysis. Yet, if the medical department was to be believed then there would not be any long-term complications except in cases which were seriously affected that night!

It was on December 9, 1984, that the Director of Health Services, Dr. M. N. Nagu, on instructions from the Union Health Minister, issued a directive stating that "under no circumstances should Sodium Thiosulphate be given in cases of MIC poisoning."

On December 11, 1984, a meeting of experts was held under the chairmanship of Dr. M. D. Saigal, Additional Director-General of Health, Government of India, which was attended also by the West German scientist, Dr. Daunderer, and Luthra, Deputy Director-General, Indian Council of Medical Research (ICMR). The Chairman announced a consensus endorsing the line of treatment adopted under the difficult circumstances at Bhopal.

It was also categorically stated that only dispossible syringes and needles were to be used. Every patient was to have been given at least five injections. Since several thousands were affected, the total number of syringes required were at least four lakhs. But did the State Health Department acquire even a fraction of the required syringes?

It was only on April 4, 1985, that the ICMR recommended the use of Sodium Thiosulphate—five months after the tragedy. The Director, Dr. Nagu, posed the ICMR with the problem of monitoring the blood, urine and chest pathology and radiology of every patient, pointing out the non-availability of syringes and funds for the purposes. The ICMR then sent another circular stating that the injections should be given to every patient complaining of respiratory complications or increased debility or fatigue. This time, the ICMR was silent over the conditions previously mentioned.

According to Dr. Sujit Das, very few noticed the extraordinary nature of the Supreme Court's interim order on the petition of the 'Janasvasthya Samity' of Bhopal on August 13, 1985 when the court directed the Madhya Pradesh to submit a scheme for the distribution of Sodium Thiosulphate (NaTS) for the detoxification of the poisoned citizens of Bhopal. The court said[1]: The scheme should also contain a provision for the distribution of this medicine through voluntary agencies and may also contain whether a health card should be issued to the victims to whom this medicine is being administered, either by the agency of the government or by the voluntary agencies."

Issuing specific order to supply NaTS to the Janasvasthya Samity, the court further stated on August 26, 1985: "We would like the state government to inform the court as to what time-bound programme the state government proposes to set up for the purposes of administration of Sodium Thiosulphate injection, with a view to detoxification of persons who are the victims of the gas. . . . We hope and trust the state government will indent the necessary quantity of Sodium Thiosulphate injections so as to meet the needs of all the persons who are affected by the gas tragedy."

NaTS is the only safe and effective antidote to chronic cyanide poisoning. Within a week of the gas disaster, Dr. Heeresh Chandra, head of the department of Forensic and state Medicine, Gandhi Medical College, Bhopal, detected evidence of cyanide poisoning in the dead bodies. His findings were corroborated by the German Clinical toxicologist, Dr. Max Daunderer and on their advice NaTS were given to the hospital patients. But for undisclosed reasons, it was stopped

1. Economic and Political Weekly, October 5, 1985.

after a few days and Dr. Daunderer was persuaded to leave Bhopal. Dr. Daunderer, before leaving India, met the officials of the Union Health Ministry and placed all the evidence before them. A joint secretary of the ministry wrote to the government of MP for consideration of the evidence and keeping the ministry informed about the course of action the government decided to take. But things did not move.

Thereafter Indian Council of Medical Research (ICMR), after double-blind control trial on the gas victims, reported: "In conclusion, the toxicological studies carried out so far have ruled out the possibility of carbon monoxide poisoning. They have clearly shown that at least in the survivors there is evidence of chronic cyanide poisoning operating as a result of either inhalation of hydrocyanic acid or more probably subsequent generation of cyanide radical from the cyanogen pool in gas afflicted victims The changes in urinary thiocyanate and in blood gas before and after thiosulphate treatment substantiated the above findings."

The ICMR recommended measures for mass detoxification of the gas-affected people living within two kilometres of the factory with NaTS and published a precise schedule for such treatment. Again nothing happened, Dr. Das retorted.

Meanwhile, the gas victims were struggling for their legitimate demands through their mass organisation, the Zahrelli Gas Kand Sangharsh Morcha. Undeterred by the violent repressive actions of the MP police, the ZGKS Morcha organised an All India Convention in Bhopal on February 17-18, 1985 to discuss "people's right to information and people's right to safe environment" with a view to organising a people's movement to achieve these rights in the context of the Bhopal tragedy. The convention participants in which included civil rights groups, science action groups, health action groups, trade unions, women's organisations, political mass fronts, etc. from all over India, constituted a Rashtriya Abhiyan Samiti and demanded, *inter alia* mass detoxification by NaTS, free medical provision for termination of pregnancy for the pregnant women who had inhaled gas in the first trimester and opted for abortion, mass campaign on behalf of the government to let the people know the possible dangers of conception before detoxification that might cripple the yet unborn and making the means of contraception freely available. As none of the demands were heeded to, the ZGKS Morcha stepped up its campaign for the implementation of the ICMR's recommendation and in April, 1985 the MP government relented, declaring that the government had already acquired 30,000 ampoules of NaTS and would soon launch mass detoxification treatment.

Again nothing happened, according to Dr. Das. The Morcha then organised the Janasvasthya Samity with the help of other local organisations and opened a Janasvasthya Kendra on June 3, 1985 on the tennis court inside the Union Carbide campus. Drug Action Forum, West Bengal, responded to the Morcha's call by sending batches of doctors periodically to Bhopal. The MP government supplied NaTS and disposable syringes. Detoxification treatment according to the ICMR schedule, brought remarkable recovery to five per cent, an substantial recovery to 70 per cent patients. The clinical and scientific proofs of chronic cyanide poisoning were found and documented. The initial positively beneficial result of the NaTS treatment brought a message of hope to about a lakh of victims who had been subjected to all forms of treatment given earlier by various agencies and the demand for NaTS treatment increased day by day. But on June 24, 1985, midnight, the MP police arrested the doctors and confiscated the Janasvasthya Kendra along with all its records and equipment, and later the MP government persistently refused to resume supply of NaTS which was not available in the market. Hence the move to the Supreme Court of India by the Janasvasthya Samity, as reported by Dr. Das.

This appeared to be a unique case in the sense that in the brief history of public interest litigation in India, no authority had hitherto been pressed to provide for specific medical relief to the community. Medical relief had so far remained a matter of charity or welfare activity, though health action groups had so long been demanding medicare for the people as a fundamental right, citing the provisions contained in the Directive Principles of the constitution. The order of the Supreme Court opened up the possibility in the jurisdical field to demand life-saving medicare for indigent people to prevent preventable death. In any case, the prospect of the victims receiving adequate monetary compensation appeared to be extremely remote. The latest reports from the Court of John Keenan, the district judge of New York, indicated that just the initial procedural formalities would take two to three years to settle and it could be anybody's guess when the actual litigation would be over.

As the plight of the gas victims is drawing more and more attention of the people in other states of India and numerous action groups are taking up the issue of destruction of natural resources by the lopsided development programmes, the authorities feel that more attention should be paid to the gas victims. The ICMR has taken up about 20 research projects mainly to study the long-term effects of the gas on human beings and the environment. These projects have been criticised by science workers as being unduly academic which will help little to tackle immediate problems. The Government of MP, in a press release dated August 25, announced a grandiose project, amounting to expenditure of Rs. 322 crore, for the rehabilitation of the gas victims and development of the disaster-affected locality. But this scheme also appears to be another deceitful manoeuvre. Out of the Rs. 322 crore, only Rs. 38 crore has actually been earmarked for the gas victims. The rest will be spent for beautification of Bhopal city, *e.g.*, conservation of natural landscape and historical-archaeological areas, development of Botanical garden, parks, playground, lake, boating and swimming facilities, etc. Also included are construction of modern airport, modernised urban transport, modernised railway station and what not. Unabashedly, chief minister Motilal Vora did not hesitate to hail this scheme as a rehabilitation project for the gas victims. Keeping in view the conduct of the state government since the disaster, one would not be surprised if this total amount of Rs. 322 crore, according to the provisions of the Act of Parliament refered to above, is later deducted from the compensation money of the victims, if received at all.

The Rashtriya Abhiyan Samiti (RAS), meeting on August 30-31 and September 1 at Bhopal, has called this so-called rehabilitation scheme a cruel joke on the gas victims and concluded that unless an agitational campaign is developed at the national level, the legitimate demands of the gas victims will remain unrealised. The most frustrating situation, the RAS has noted, is the peculiar apathy of the opposition political parties and their mass fronts, particularly the leftists, to concern themselves with the Bhopal disaster. It appears that everyone is afraid to cause annoyance to a big multinational. A curious incident will be self-explanatory. A few weeks ago news reports said that Sivraj Patil, minister of state, declared in the Lok Sabha, while replying to a question, that the licence of the mysterious R and D Centre of Union Carbide at Bhopal had not been renewed. This statement appeared to be a clever distortion of facts. For R and D activities no licence is required. The R and D laboratory of Union Carbide, had a certificate of exemption from income tax and presumably that is what has been discontinued. It has not been debarred from carrying on the so-called R and D activities. A representative of the ZGKS Morcha went to a leftist MP (from West Bengal) and, after furnishing documentary evidence, urged him to move a motion against the minister accusing him of misleading Parliament. Strangely, the leftist MP defended Union Carbide, stressed that Parliament should not be bothered with trivial matters and advised the Morcha not to overestimate the importance of the Bhopal disaster.

Given this situation, the RAS has appealed to all concerned people to organise in defence of the gas victims. A national programme of protest has been adopted. A Bhopal Week from November 27 to December 3 will be observed all over India. Conventions and rallies were held in all states demanding necessary medical relief, rehabilitation and compensation for the gas victims. After collecting signature from 10 lakh people, these demands will be submitted to the Prime Minister at the venue of the Indian Science Congress which met on January 3, 1986. A National Seminar was held at Delhi on January 4, 1986 on the Bhopal disaster. In West Bengal a 'No More

Bhopal Committee' had been constituted comprising a large number of organisations to implement this national programme of protest and to campaign for the people's right to information and to safe environment.

What was the attitude of the state administration? It was evident from the fate of the judicial inquiry being conducted by Justice N.K. Singh of the MP High Court, that utterly disgusted with the goings-on, Justice Singh flayed the State Government for its non-cooperative attitude.

Two noted film producers—Tapan Bose and Suhasini Mualy —who shot some 250 minutes of 16 mm film—the world without tomorrow—formed a citizens' relief and rehabilitation committee. According to a correspondent report these two independent film makers launched into a tirade: "There is a massive cover up. The government does not want people to be treated as MIC victims. The confusion was understandable in the first week after the tragedy, but even now, they are only using bronchodilators, cortizene and corticosteriods—nowhere else in the world are they dispended in such quantity. These have dangerous side effects—steroids and cortizenes pep you up and encourage you have recovered."

Bose, Mulay and their team contacted local lecturers and businessmen to donate cash and food, which they distributed selectively since December 20, 1984 to those without ration cards. While the state government proclaimed that rations were given to the affected, free of cost, a number of victims did not have death certificates for the deceased were running from pillar to the post to obtain Rs. 10,000 compensation offered by the state government:

> "The local administration simply hasn't grasped the enormity of the situation," Bose concluded.

Much of the official relief was being channelled through the local Congress (I) MLAs and Councillors and was not reaching those it was intended for. The state government announced ex-gratia relief of Rs. 10,000 to the dependents of each dead person: if the beneficiary is a child, the amount is held in a bond in the child's name by the Collector. The 'seriously afflicted' were given Rs. 1,000 to Rs. 2,000; while the 'moderately afflicted' could receive Rs. 100 to Rs. 1,000. A month after the tragedy Rs. 36 lakhs were distributed in this manner, and an almost equal amount given to the railways for its affected employees. But the mood of the killer gas victims was defiant:

> "It's all over for me anyway. I sold our gold for Rs. 400 a tola (a fifth of the market price). Whether I get Rs. 10,000 or Rs. 20,000, will Union Carbide or the government replace my dead son?"

No one had really fathomed the economic dilemma of the victims. Mohammad Bashir, who lost his wife and two sons, told that how he reported back to work at the Bhopal Textile Mill but felt exhausted after three hours. The supervisor advised him to take a breather for 10 minutes but he was not able to continue:

> "All the other workers who had inhaled the gas were in the same state. None of them, even from Union Carbide knew what wages they would receive—if any at all—while they were absent. They also spent a fortune on treatment at private clinics selling jewellery—Marwaris at the Saraph ran out money to pay for ornaments being dumped—even utensils. Many of the colony dwellers were porters, loading 60 kg sacks unto tracks, which they simply could not do now. Construction activity grounded to a halt." A telling epitaph to the much vaunted "Operation Faith" was a run on locks, when over two lakh—of the 8 lakh residents—fled the city; a single padlock went for Rs. 100 on December 12, 1984, about 50 men, women and children, all victims of the recent poisonous gas leakage tragedy, demonstrated before U.S. Embassy, New Delhi, demanding the closure of all such factories in the world. Carrying placards, the processions started from Jama Masjid where the victims had been camping since their arrival in Delhi. Later, a memorandum addressed to the US President was handed over to the embassy officials.

The memorandum in Hindi demanded that the Union Carbide factory in Bhopal be closed down for good and all other factories manufacturing such poisonous gases be banned immediately to prevent a similar tragedy elsewhere.

The victims also demanded a compensation of Rs. 10 lakhs each to the families of those killed and Rs. 1 lakh each to the families who were affected by the leakage of the poisonous gas. According to American experts, the people who received relatively light doses of the gas and who had, therefore, survived so far could die as a result of ordinary respiratory infection. "Their lung tissues are damaged by the gas and even a common respiratory infection could become fatal in their case."

Loboratory tests had shown that animals inhaling the methyl isocyanate in heavy concentrations experience an immediate swelling of the tissues that line the air passages leading to the lungs. The swelling could be sufficient to close the air passages. Victims who survived, this primary effect suffer massive lung damage that caused fluids to be secreted in the lungs and as a result, in a day or so, the victim in effect drowned in his own fluids and died.

Laboratory animals exposed to lower doses of isocyanate escape these primary effects but they still suffered serious tissue damage within their lungs and this was what made them particularly vulnerable to ordinary respiratory infections which could be common in poor hygienic of living. The people who had already died were, therefore, not the only vi tims of the gas leakage frcm the Union Carbide plant.

There could be thousands of people living in Bhopal today whose lung tissues could have been damaged by the gas and who could lose their lives in the next few weeks or months, or years. The gas leakage disaster in Bhopal had, therefore, much deeper dimensions.

Here in the United States, the Bhopal tragedy had created a a scare. It had raised the question whether the storage of poisonous gases in underground tanks in this country was safe. Was there any law that regulated the storage of such lethal gases in underground tanks? Who enforced that law, a government agency like the environmental protection organisation or the industrial management that stored the gas?

Inquiries should that, as of today, there was no law in the United States that would empower any government agency to inspect or regulate the storage of poisonous gases in underground tanks. The occupational safety hazards agency took care of only the health and safety of the industrial workers within the premises of their work areas against the hazards of leakages of gas or use of poisonous chemicals. It did not concern itself with the impact of any gas leakage on people living outside the factory premises.

The ony other government agency that could be concerned with the industrial leakages of lethal gases from factories and storage tanks was the environmental protection agency (EPA). But it did not have any powers to inspect or regulate the industry's safety devices.

4

BHOPAL'S AGONY UNABATED

A POTENT combination of highly toxic gas leakage, particularly of methyl isocyanate surged out from the Union Carbide plant claimed officially 2,500 lives and affected thousands of others in various other ways under the presence of a runaway chemical reaction. The plume of poisonous gas, about 43 tonnes of it leaked out in less than two hours, soon enveloped over half the city of Bhopal. By the time it settled, more than 500 people dropped dead. Before the week ended a conservative estimate of 2,500 people died from excruciatingly painful chemical deaths from swollen and flooded lungs, asphyxiation, severe brain damage, broken nervous system and charred indides. So far neither the government nor the Union Carbide management provided accurate information to the gas victims about the poisonous gas that leaked out on December 2/3, 1984. This information was vital for devising comprehensive treatment of the gas affected people. No wonder, doctors came out with various theories and modes of treatment.

Even before the dead count was half way through, Union Carbide's senior executives in the US declared that "MIC is not fatal; it is like a potent tear gas", and that the worst already had been done. Union Carbide India Limited (UCIL)'s officials had earlier refused to divulge the identity of the chemical that leaked and failed to sound the alarm on time to warn the people. Panic gripped those who woke up choking, breathless with their eyes burning. Between three and four lakh people were exposed to MIC and other poisonous gases for varying periods of time. That was the beginning of the world's worst disaster, the most destructive episode in its industrial history.

In the absence of any open toxicological literature about this fatal gas the local doctors were also initially taken by surprise. No confirmation was given by the UCC authorities or the UCIL management about the nature of the poisonous gases.

A raging controversy developed around the administration of sodium thiosulphate injection, usually considered an antidote for cyanide poisoning. The matter went right upto the Supreme Court which directed the state government to make sodium thiosulphate injections available to all those who wanted it, and widely publicise where the injections would be administered. In fact, facilities available at that point of time to meet the requirements of the enormous number of gas victims who were seriously affected, especially those who had suffered damage to their respiratory system due to the inhalation of the gas were terribly inadequate. The end was not in sight. About 1,00,000 people continued to suffer from varying kinds and degrees of medical disability directly attributable to their exposure to chemical poisons. Between 30,000 and 40,000 of them had serious ailments. Many never would recover. So the tragedy did not end there. It persisted and aggravated day after day. For the majority of those who had been suffering still did not know what was the proper medical care and how to get it.

Soon after, in the latter half of December, 1984 when the patients started returning home and required after care as out patients, special MIC clinics were set up in the Medical College in the department of medicine, Opthalmology, Paediatrics and Gynaecology. These were assisted by other investigative departments like Radiology and Pathology. The six community medical centres in the gas affected *basti* of Bhopal and the tertiary care centre set up at the Hamidia Hospital, the city's biggest catered to the health needs of less than half of those who needed medical aid on an urgent and continuing basis. Proper medical records detailing the changes—often deteriorating health status of the gas victims—existed in respect of less than 10,000 patients.

A special MIC ward was started on February 1, 1985 in the Medical College. It had 60 beds for indoor care of seriously ill MIC patients who were investigated in depth and whose medical care was designed. Besides, it had various diagnostic aids available on the same floor, ensuring that the treatment was scientific and well-controlled. The treatment continued to be haphazard and the medical community in Bhopal divided over the therapeutic approach to be followed. Both over and under medication persisted. Even symptomatic treatment was difficult in the absence of adequate information about precisely what chemicals leaked and how they affected the body in the short and the long run.

HEALTH HAZARDS

The killer gas said to be methyl isocyanate (MIC) which escaped on the fateful night, killed several thousands and crippled many more. The manner in which deaths occurred left little doubt that they were due to acute asphyxia *i.e.* acute lack of oxygen in the body. The latter could have resulted from either acute damage interfering with oxygen-carbondioxide exchange or from poisoning of the oxygen carrier haemoglobin preventing the release of oxygen in the tissues for metabolic activity. Possibly both these mechanisms were at work.

Thousands of survivors in the acute phase suffered from severe symptoms linked to respiratory, gastro-intestinal opthalmic musculoskeletal systems. In addition, the psychic trauma caused by the tragedy with the resultant feelings of anxiety, insecurity and deppression only made the situation worse. This not only generated new symptoms of varying nature but also intensified those due to organic damage. Of the various organs affected by the MIC, the respiratory system has been the worst affected. Most of the affected population suffered from cough with or without expectoration, distressing breathlessness especially on exertion, inability to work and excessive muscle fatigue—apparently all attributable to lungs.

In order to assess the extent of the damage to the lungs, it was necessary to find out the effect of gas inhalation on the lungs. The process of investigation is a complicated one which gives a precise idea of the state of lungs, which can not otherwise be derived by routine chest radiology. At the time of the accident, the Medical College had a conventional apparatus to assess pulmonary function, which could not meet the demand. The anxiety of the local medical personnel was resolved by the spontaneous help extended by Prof. A.S. Paintal, director of V.P. Chest Institute, Delhi, who not only lent his equipment to the local medical college to meet the urgent need but also placed at its disposal the service of Dr. S.K. Jain, professor of the institute and an expert in testing pulmonary functions. This led to the establishment of one of the most sophisticated computerised pulmonary function testing laboratory in Madhya Pradesh. It had the capability of evaluating functions of the lungs both in a steady state and during exercise, enabling the various parameters to be worked out with precision and in great detail.

This data could be stored in the computer attached to the equipment and could be analysed fcr comparison at a later date. This data could be re-called and assessed for each patient, besides getting a printout for record. This modern laboratory increased immensely for the capability to handle complicated cases. The facility for the study of blood gases and PH of blood, which gave the content of oxygen and carbondioxide in both arterial and veinous blood was also available. To look into the lungs and air passages, the facility of bronchoscopy had been provided, which also helped in detailed study of air passages by broncho-alveolar lavage and biopsy. It gave the exact idea and state of disease in the lungs.

Lungs biopsy revealed that the gas victims developed two types of changes:

> One group had Brochiolities obliterans and the other Interstitial fibrosis, which limited the capability of the lungs to exchange gases and thus resulted in breathlessness. The patients who had fibresis of lungs were likely to suffer from the symptoms of breathlessness. They would require constant medical supervision for many years. The main object of carrying out these detailed investigations was to detect

the type of pathology that an affected developed and provided him effective treatment. The result of the treatment could also be evaluated by carrying out specific tests.

There were broadly two types of abnormalities in pulmonary functions:

Obstructive—which indicated narrowing of obstruction of airways—and restrictive—in which the lung lost the normal power of expansion and relaxation, and thus could not exchange gases normally. The latest equipment installed at the MIC ward had the capacity to give details of these abnormalities, recorded them on the computer and compared changes in these functions after treatment, guiding the doctors in future management of these cases.

Needless to say, the patients had been treated on a scientific basis. Patients with obstructive illness had been given bronchodilators and steroids, if required; and those with restrictive disease had been treated with steroids alone. There were cases which were being treated with sodium thiosulphate to detoxify them from toxic effects of the gas.

MIC ward had 60 beds and various laboratories which formed the local point for all research activities being conducted by Indian Council of Medical Research to find out the manifestations of the disease caused by inhalation of toxic gas and to derive the best method of treating these patients. Various experts from India and abroad visited the ward and gave their valuable advice. Dr. A.J. Newman Taylor came from England in the month of March, 1986 and studied some cases. Earlier, experts from the United States had come in a group. In fact, experts from all over the country have been frequently visiting the ward to help in the treatment of the cases. Thus, the ward became a "lighthouse" in the ocean of unknown maladies caused by MIC. The concerned effort by the medical profession in this highly specialised ward with all modern facilities would go a long way in alleviating the suffering of gas victims and providing them a new ray of hope for life. Most of the affected population suffered from troublesome cough with or without expectoration, disturbing breathlessness especially on exertion, inability to work and excessive muscle fatigue—apparently all attributable to lungs. The questions immediately asked were:

1. What is the nature and extent of damage to lungs?
2. Is the damage to lungs permanent or will it revert to normal-spontaneously or following treatment?
3. What is the best way to treat these patients?
4. What complications could possibly occur in the future?
5. Will the disease caused by MIC be progressive?

The answers to the above questions were considered important for treating as well as protecting the affected population from the disabling effects of MIC lung disease. Thus, an accurate assessment of the patients' conditions was considered to be paramount importance. In order to achieve this, sophisticated equipment was urgently needed to measure/estimate the effects of MIC, on static and dynamic lung volumes, mechanical efficiency of the lungs, oxygen-carbondioxide gas transfer mechanisms, quantity of oxygen, carbondioxide and acid base ion in the arterial blood, pressure and flow rates generated in airways, heart and pulmonary circulation and above all responses of heart circulation and breathing to physical exertion or exercise.

Unfortunately, facilities to carry out all these investigations were not available in Bhopal at the time of the tragedy. In fact, it was not till the beginning of January 1985 that some simple pieces of equipment such as blood gas analyser apparatus could be made available. The information collected with the help of these instruments proved very useful in defining the pattern of MIC lung disease as well as in the management of the suffering patients. However, the need for more sophisticated equipment was felt acutely. The equipment was eventually provided in March 1985 by Prof. A.S. Paintal, head of the centre for the study of visceral mechanisms of department of science and technology at the V.P. chest institute, University of

Delhi on loan to Bhopal. Work was immediately started in collaboration with Prof. N.P. Mishra, head of the department of medicine, Gandhi Medical College, Bhopal. The problem to be tackled was of gigantic proportion. The exact number of sufferers was known only after epidemiological survey of lung cases. The equipment that was acquired was fully computerised. Such an equipment offered many advantages;

(i) a large number of tests could be carried out at a speed of 20-30 times faster than was possible manually;

(ii) the measurements were more precise and accurate;

(iii) the results were available instantly;

(iv) repeated observations could be automatically compared; and

(v) the data could be stored indefinitely and retrieved whenever required.

Thus the need for thorough investigation of these patients and a prolonged surveillance for at least a decade was possible with the help of such an equipment. The first task was to track down those cases who showed definite evidence of lung involvement in the acute phase following the MIC exposure. The lesions were either detected at that time by the chest radiographs or by expert physical examination by consultants of the hospital. Also included were patients in whose family or immediate neighbourhood a death attributable to MIC had occurred or in whom there was circumstantial evidence of a heavy exposure to MIC. A detailed study of such cases repeated periodically helped in answering most of the important questions raised above in connection with the ultimate fate of the lung disease caused by MIC.

These patients were brought to the laboratory for investigations. They were made to blow into the volume and flow measuring instruments several times. The computer system examined the various breaths and measured the actual expansible volume of lungs, the various flow rates generated in the lungs and different breathing capacities. Another important test carried out was whether the gas exchange for oxygen in the lungs was normal or had been adversely affected by the MIC. Blood from the artery i.e. purified blood from lungs was collected and oxygen, carbondioxide and acid-base ions (PH) were estimated.

Most important of all was the study of response of the heart and lungs to physical exercise. For this purpose, the patient was made to ride a stationary bicycle which was electrically braked. With time, the brake became harder and harder so that while pedalling, the patient had to do a constantly increasing work (physical exercise). Simultaneously, the patients' breath, heart rate and ECG were recorded and also his breath was continuously analysed for oxygen and carbondioxide. The computer on 'line' analyses all these signals and printed eighteen different parameters continuously every fifteen seconds throughout the period of exercise. Important amongst these parameters were heart rate, respiratory rate, size of breath, minute ventilation, energy consumed, oxygen intake and carbondioxide output. These measurements provided valuable information regarding the cause of breathlessness and fatigue.

The work was in progress and 238 patients had already been examined—some of them repeatedly. Some of these patients had been seen earlier in the month of January-March, 1985. The findings revealed that MIC produced lesions mostly in the smaller air passages were also in patches destroyed the alvcoli. Lung biospies had also been done in selected areas. There was evidence that in most patients smaller air passages had been inflamed and destroyed and in the other this was combined with destruction of alvcoli also. It was not commonly understood that the destroyed tissues—converted to scars—could not be regenerated. A vast majority of the patients were emotionally disturbed which was definitely making the existing symptoms worse and also generating new symptoms. These patients were in urgent need of rehabilitation—physical as well as emotional.

They are being given diversional therapy and occupational therapy. They also need routine care of their chest problems by physiotherapy, breathing exercises, gradually increasing physical exercise and drugs as may be indicated. The possibility of a bio-chemical poisoning of the cells is being kept in mind. Only further researches and follow-up of these patients will help in unravelling the mystery of the MIC lung disease. A long term surveillance of these patients is proposed to be carried with the sophisticated equipment for lung function testing and blood gas analysis.

It was clear even in the early stages of clinical management of the patients that the health infrastructure should be adequately strengthened and expanded to meet not only the short term, more so the long term needs. Currently medical services to the patients exposed to the toxic gas are being provided by a string of dispendaries, special clinical examination-cum-treatment centres and hospitals. The Health Department has set up a 3-bed hospital in the affected area with all necessary equipment and laboratory facilities. Medical specialists, gynaecologists, ophthalmologists, surgeons, paediatricians, pathologists and radiologists have been stationed in the hospital.

Sixty beds are reserved in the Medical College exclusively for the gas victims. There are 20 beds available in special eye hospital opened with the assistance of the Royal Commonwealth Society for the blind. Special clinics have been set up at 17 places in the 30 affected municipal wards. Some of these are equipped with X-ray and laboratory facilities. The 30-bed hospital in J.P. Nagar, one of the worst affected areas, has been expanded by building a permanent shed with 12 rooms and spacious waiting room to serve as an out-patients department. Indian Council of Medical Research is also running a laboratory in this hospital and equipment for blood-gas analysis, urine thiocyanate test and double blind study is available here. Among the dispensaries are five polyclinics were X-ray machines are available and specialists are posted for attending to the patients. Six dispensaries have well equipped laboratories. The Medical College ward is equipped with the most modern facilities, some of which are not available anywhere else in the country.

It is equipped with a computerised blood-gas analyser to study the capacity of lungs to oxiginate blood and discharge carbon-dioxide.

There are 9 ambulances reserved exclusively for the gas affected areas and in addition there are 3 mobile medical vans for these areas. The hospitals and dispensaries have been given sufficient stock of necessary medicines so that the people do not have to go to private doctors for treatment. Those who have suffered the impairment of their vision have been given free spectacles. The clinics and dispensaries have been provided with health cards for systematically recording the treatment given, in order to facilitate the following action. Special documentation is being done for indoor patients and out-door patients who were exposed to the toxic gas and need long term treatment and follow-up. Special attention is being paid to pregnant women with regard to ante-natal care. The Indian Council of Medical Research has taken up a project for the follow-up of all pregnant mothers and the new born in a population of 88,000 in 15 wards. It is proposed to follow-up the growth—both mental and physical—of these children for a period of 6 months.

A psychiatric clinic has also been set up with the help of experts from Lucknow Medical College. More than 1000 cases have so far been registered here. The Indian Council of Medical Research with the collaboration of the Medical College has also taken up an epidemological survey of 17,000 families covering 85,000 persons in 13 affected wards. All these families are being followed up through fortnightly visits made by the medical and para-medical staff. Those still found to be suffering from the after effects of the toxic gas are provided treatment, at home or at the clinic or hospital as the case may be. In addition to this, 20 research projects and studies have been taken up in collaboration with the Indian Council of Medical Research to assess the various possible effects of MIC in human-beings.

Although detailed documentation about the ailments is available from the socio-economic survey as well as from the medical records maintained at the dispensaries, clinics and hospitals, it has been decided to carry out a fresh survey at this stage through house to house visits by medical teams. A team of 60 doctors has been deployed for this job. The doctors provide referral slips to such of the patients as are not being treated at the clinics and direct the patients to report at the nearest clinic or the hospital for proper treatment and follow up. Since the families taken up in the ICMR epidemological survey are being followed up by the Medical College teams they are not covered by these doctors because it would lead to duplication.

In addition to these, several voluntary organisations have come forward for conducting such surveys. Recently, a team of 60 doctors from Raipur came to Bhopal under the auspices of the Red Cross Society, Raipur Branch and Akhil Bhartiya Rachnatmak Samaj, Raipur. There are several members of legislative assembly in Bhopal who are qualified medical doctors. Four such doctors who are members of the Legislative Assembly also went round the affected areas and assessed the health status of the gas victims. The symptoms of tuberculosis among some patients were aggravated after exposure to the gas. Special efforts are being made to detect such cases and provide proper treatment to them. The areas, which were affected, are mostly inhabited by poor people. The exposure to the gas has weakened them. The Government has, therefore, started a mass immunisation programme for children in this area. Apart from medical treatment it is also proposed to start shortly two Yoga centres in this area for the treatment of lungs by teaching breathing exercises.

A well-equipped clinic has been established for long follow-up of patients to detect any possible late effects of MIC on eyes. The clinic is equipped with modern instruments such as general Specular Microscope, Slit Lamps Fundus Camera and Ophthalmoscopes for the study of the anterior and posterior parts of the eye. ICMR has sponsored and financed a study for long term follow up of the occular changes in gas affected cases. Instruments such as electro-retinogram and electro-occulogram are being procured for further studies. More than 3,000 patients reporting everyday at every clinic and dispensary. It is clear that the health facilities available in the affected areas will not be able to provide all the medical support necessary for the affected. These injections and have qualified doctors and a hygenic place for setting up a clinic are also being supplied these injections free of cost. Injections have also been supplied to IMA.

Contrary to Union Carbide's claims, the health damage weakened by methyl isocyanate and associated gases released from the company's Bhopal plant about two years ago remains widespread and severe. Evidence collected from numerous medical studies, including the 23 projects sponsored by the Indian Council of Medical Research, proves that MIC is a far more potent toxin than it was originally believed to be, that the chemicals released during the episode have caused extensive damage to a number of organs, and systems in the body, and that some of the damage is basically of an irreversible kind.

Fully more than one year after the catastrophe, about half of the victims who underwent moderate or severe exposure remain debilitated. Nearly 1,00,000 people are unable to work or live normally. And about 50,000 have shown a significant deterioration in their condition, akin to those who are afflicted by a degenerative disease. The response of the gas victims to medical treatment has been far from homogeneous or positive. In about a fifth of them it is close to nil. Thus the future of tens of thousands of them is bleak, the prognosis for many deppressingly poor. The most widespread disability found among the Bhopal gas victims pertains to the lung. Various studies conducted over the past years have established that MIC and associate gases destroyed the air so badly as to repair the functioning of the vital organ in a majority cases. The incidence of abnormal lung function and a greatly lowered vital lung capacity remains widespread.

Thus half the exposed population can not breathe normally; many get acutely breathless on walking in a few steps or even at rest; and few can do any kind of manual labour, even of the soft variety. The lung tissues that were destroyed can not be regenerated. And in some cases, fibrosis—a galloping disease that gradually solidifies and eventually kills the lung—has already reached an advanced stage. The response of patients to drugs such as broncho-dilators, cortico-steroids and to respiratory physio-therapy has been uncertain in no small proportion of cases. The primary cause for this, it is now known, is that the victims are affected by chronic immunological lung disease—a condition that has worried the population. Steps have, therefore, been taken to strengthen the medical infrastructure. A new 100-bed hospital at a cost of Rs. 2.5 crores is already under construction. This will be equipped with the most modern and sophisticated equipment. The capacity in the Medical College is also being expanded. Modern facilities are being introduced in the Kamla Nehru Women's Hospital for the care of mothers and infants.

Sub-acute syndromes of MIC toxicity have now become evident among those who had been most exposed but had survived. Although the main symptoms still relate to the respiratory tract, symptoms can be grouped into the following main areas. These are still the main group. Typical symptoms are breathlessness and quick fatigue. Some patients have responded well to treatment whereas others have not responded so well. Detailed clinical, biochemical and radiological studies, including pulmonary functions, blood gas analysis and pathology by open lung biopsy have been carried out on these patients. The patients have developed fibrosis of lungs which affected their breathing capacity.

Studies on the administration of sodium thiosulphate have been advised by the experts, particularly about pulmonary functions after the injections. So far sodium thiosulphate injections have been administered in accordance with the guidelines issued by the Indian Council of Medical Research on the basis of extensive discussions in a meeting on April 4, 1985. Eyes of about 2,00,000 individuals were affected in varying degrees depending on the duration of exposure. There has been no reported case of blindness. However, there are still widespread complaints regarding watering of the eyes and 'foreign body' sensation. Cases of deterioration in vision are being attended to and spectacles provided where necessary. The symptoms in gastro-intestinal cases are lack of appetite, burning sensation, etc. Endoscopic examination of such patients has revealed the presence of gastritis, oesophagitis and ulcers in some cases. These patients are being treated on the usual lines and have shown improvement. There are voluntary agencies also which are actively supplementing the efforts of the government in providing medical care to the patients. All help is being given to these agencies and medicines are being given free to such agencies. Sodium Thiosulphate injections are being administered according to the guidelines given by the Indian Council of Medical Research. Such voluntary agencies which are desirous of administering puzzled specialists, especially because in most cases it has not been caused by repeated exposures to poisonous MIC over a period of time. Indeed the most disturbing recently discovered to be a poison with long-acting effect, is that it has inflicted a serious damage upon the immune system of the body. A study sponsored by the ICMR, and conducted by the KEM Hospitals and the Cancer Research Institute of Bombay with assistance from the School of Environmental Health, University of Pittsburg. USA has produced some vital and definitive insights into the way MIC works.

It has found specific MIC antibodies (developed by the body in defence) in over three-fifths of all exposed subjects, thus demonstrating that the most important casual agent at work in producing chronic pathophysiological changes in the victims was MIC, and not cyanide, as suggested by some hypotheses. The study has shown that MIC exposure is correlated with the increased level of an immunoglobulin IgE, which is involved in hypersensitivity/allergy mediated abnormal body response. It has also found that MIC lowers the response of certain specific cells, T and B cells, and their subsets, which are crucial to the body's immune mechanism for fighting disease. This means, very simply, that MIC has damaged the natural immune system, and hence that its victims are now both more vulnerable to all manner of infections and less likely to be cured of the ailments, afflicting them, especially those of the lung. The lungs are not the only site of health damage. MIC and associated gases have proved destructive of a range of soft tissues in many parts of the body. They have damaged the eyes the central nervous and neuro-muscular system and the brain. There are cases of damage to the gastro-intestinal tract and some abdominal organs such as the pancreas, too. The chemicals are strongly suspected to be toxic to genetic cells.

Experiments in a number of laboratories suggest that MIC may produce cancer and genetic damage. And last but not the least, the chemical poisons have produced profound neurological and psychiatric disturbances and mental illness, especially amongst children. More than 10 per cent of the adults examinep had neurological complaints in the acute phase such as hearing loss, tremor, vertigo and muscular weakness. A preliminary survey of 205 school children in Bhopal shows that six per cent have epilepsy and as many as 68 per cent have had difficulty in keeping up with their studies over the past year. The ICMR sponsored pregnancy outcome survey, covering a total of 2,500 women known to be pregnant exactly at the time of the catastrophe, shows that the rate of spontaneous abortion in the first 20 weeks of pregnancy was 14.8 per cent, more than twice the normal rate of 7 per cent recorded in various parts of the country. The evidence in respect of psychiatric problems is even more disturbing. Approximately 22 per cent of those screened among the people reporting at various government clinics in Bhopal were found to have mental disorders. As many as 37.3 per cent of them were afflicted by neurotic deppression, 24.9 per cent by anxiety state and 35.2 per cent suffered from adjustment reaction.

According to the recent sample of cases detected among 276 of a total of 700 randomly selected households shows that 72 per cent suffered from neurotic deppression and 27 per cent had anxiety state. The number of people who are yet completely to overcome the trauma of what they went through a year ago or of those who still have nightmares about the episode, its immediate aftermath and the humiliation they experienced runs into tens of thousands in Bhopal. Few of them can be treated by drugs. Most, especially children who are among the worst sufferers, can only benefit by psychotherapy, sympathetic counselling and a massive effort at rehabilitation which restores in them the will to live, assures them of society's support and instils in them the hope that they can lead a meaningful life.

The difficulty in respect of treating psychiatric disorders applies to other ailments to a certain extent, too. For, MIC still remains an inadequately understood poison. Union Carbide Corporation which has more data on it than any other agency, is not forth-coming. Despite having already spent Rs. 5.5 crores on various research projects, the ICMR has not arrived at an understanding of the complexity of the problems or at a line of treatment which it would regard as satisfactory. MIC has continued to present new and baffling questions. As the research goes on—some projects will carry on for several years—it would become evident that the medical problems created by the disaster, are as great in magnitude as the tragedy itself.

A large-scale epidemological study, covering a sample of 85 000 people, and several other projects, including one for a comprehensive cancer registry, and many to study the long term effects of MIC and other gases, have now been launched. Some of them are certain to yield valuable results but that is in the future. But until they do so and until the conclusions are turned into therapies and vigorously implemented, a large number of people in Bhopal will continue to suffer for no fault of their own/than that they happened to live near a plant that was deliberately under-designed and operated in gross violations of safety norms by Union Carbide Corporation.

Medical and Ex-gratia Relief

The problems of providing relief and rehabilitation to the victims of the gas tragedy in Bhopal has two aspects, namely; (1) immediate relief and (2) long term relief and rehabilitation. In the initial months following the tragedy, the entire administrative and medical machinery was geared to the task of providing immediate medical relief and cash relief to the sufferers of the accident. After the first few months when things become systematic and the pressure was somewhat reduced, schemes for continuing medical relief and social and economic rehabilitation was launched.

Social rehabilitation schemes have been formulated with a view to providing services needed for that section of the gas-affected population which is socially deprived and has been badly affected. A socio-economic survey of the affected population, covering about 80,000 families, has been completed. The bulk of the affected population comprises people living below the poverty line. The magnitude of the problems of malnourished children, pregnant women and nursing mothers, disabled persons and persons who have contracted chronic diseases which need long term treatment, in very large.

Three integrated child development services projects have been started in the gas affected areas. 393 Anganwadis are functioning under these schemes, the schemes focus primarily on pregnant women, nursing mothers and children upto the age of 16 years. The package of services provided consists of non-formal pre-school education, immunisation, health check-ups, supplementary nutrition, medical referrel services, nutrition and health education for women. The number of beneficiaries is about 1,00,000.

Education in these areas for children is another important felt-needs. Six schools have already been started in the area in rental buildings and permanent buildings are being constructed for these schools. Apart from this, it is also proposed to set up schools which will provide board and lodging facilities for the children so that they can be properly looked after. Community centres are also proposed to be set up to provide social services to the children in the age group of 6-18 years. These centres will be the focal points where children as well as elders of the community can meet and avail themselves of the various facilities.

Preliminary studies reveal that three-fourths of the population of the gas affected areas is devoid of proper civic infrastructure. The tragedy ridden area is sub-standard from the point of view of even basic infrastructure facilities. The area is devoid of basic facilities like water supply, paved roads, toilets, drainage, open spaces etc. Stagnant water bodies and lack of sanitation are the striking problems. Therefore, it was felt that this area needed total rejuvenation so as to achieve qualitatively acceptable environmental standards and bring it to acceptable levels of living.

It is proposed to take up environment of about 40 identified slums. The improvement of will consists of providing paved approaches, service-drains, low-cost sanitation, street lighting, water taps, shelter improvement, development of open spaces, construction of drains, reclamation of low-lying areas, improvement of open wells, solid waste disposal arrangements and development of house sites. In order to prepare detailed plans for these works, 7 different task forces have been set up involving the departments of Town and Country Planning, Public Works, Public Health Engineering, Public Health, Education and the Department of Environment. These task forces submit their plans to a coordination cell which clears the schemes and puts it up to the government for their approval. Work has already started in various affected wards. The work of environmental improvement is going fast apace in J.P. Nagar colony, one of the worst affected colonies. A sum of Rs. 58 lakhs has been sanctioned for improvement work in a few other colonies while plans for the rest are being formulated.

Similarly, the Irrigation Department of the MP Government and the Public Health Engineering Department have prepared schemes for training of three major drains, namely, Patra Nala, Tila Jamalpura Nala and the Subhash Nagar Nala. Sites have been identified for 1500 houses to be built for the victims of the tragedy. The work on these has already been started. These houses have a unique feature that each house has a separate work-room where the owner of the house can set up cottage industry if he so desires. Services of *Sulabh international* have also been availed to build community toilets to begin with. When trunk sewers are laid it will be possible to connect individual toilets to these community toilets. One of the task forces prepared detailed plans for area development of these localities.

The work of relief and rehabilitation has been constantly monitored at several levels. There is a State-level Relief Committee headed by the Chief Minister to advise the government on relief and rehabilitation measures. People's representatives and public men from various fields have been included in the Committee. Then, there is a Cabinet sub-Committee under the chairmanship of the Chief Minister which consists of the finance minister, minister of health, minister of labour and industries and minister of social welfare. The Cabinet Sub-Committee meets frequently and monitors the progress of work. A separate department called the "Bhopal Gas Tragedy Relief and Rehabilitation Department" has been created in the state government to look after this work. The new department is directly responsible for the administration of relief and rehabilitation in the affected areas. At the field level the affected areas have been clubbed into a separate health district under the charge of a chief medical officer. The 40 wards, which are affected have been divided into 7 sectors Each sector is manned by officers of the District Administration and each sector officer has his office located in the area itself. The sector officer is responsible for supervising the entire gamut of relief and rehabilitation activities in his sector. The work of socio-economic and physical rehabilitation has picked up tremendous momentum and will help in transforming the lives of the people who unwittingly became victims of the world's worst industrial tragedy.

It was decided without question that an amount of rupees would be given as *ex-gratia* cash relief to the next of kin for each of the dead in a family, not as compensation for the life lost, but by way of temporary relief to help the affected family. The magnitude of the problem was so vast that the first few months were spent in providing immediate medical relief, assessing the extent of the havoc wreaked on the town and its people by sifting of enormous quantities of records and meticulous enquiries in the affected localities. Hence, payment of ex-gratia cash relief to the survivors could properly commense only from April, 1985. The officials had to reconcile the need for immediate payment and the no less important need to ensure that it is received by the genuine claimant and not an imposter.

Initially all the claimants had to go to the Collectorate to have their claim verified. Then, during one of his numerous visits to the gas affected localities, the chief minister ordered that this verification should be done at the door steps of the victims. The affected area was divided in 7 sectors, each under a sector commander who was provided with adequate staff. The people who had claims went to the sector commandant within easy reach of their homes or the sector officer could visit their homes to complete the formalities of verification.

The names with addresses and photographs where available of the dead were published through posters in Bhopal and newspapers all over the State so that their heirs, if any doubt, could come forward to make their claims. So far the ex-gratia payment of Rs. 1.4 crores has been made in 1400 out of the 1754 possible claims. Fresh lists have been published in the newspapers on the 5th October, 1985 for the remaining persons. Some of the dead may not have any survivors. But all efforts are being made to reach the legitimate claimants. J.P. Nagar, one of the worst-affected localities is a witness to the re-affirmation of life. Pucca houses are being built with the help of cash relief. The environment has improved. A widow whose husband perished in the accident has set up a vegetable shop. Another person who used to push a cart and sell fruits from door to door has bought a three wheeler auto-rickshaw. Such examples are numerous.

There were thousands who became sick and unable to do any work. Many suffered serious impairment of their physical capacity and became incapable of doing hard physical labour. There were thousands of others who were tending their sick relatives. Free rations have been distributed to the victims of the gas tragedy. In view of the circumstances, it was felt that poor people, scattered all over Bhopal, should be supplied with free ration till they could stand on their own feet once again and recoup their losses at least partially. A massive drive to prepare ration-cards was launched. In 30 municipal wards 183 Fair price shops were mobilised to distribute free rations. It was a massive operation, successfully launched. The system was started in the month of December, 1984 itself. Till the month of September, 1985 27,306 quintals of wheat and 16,189 quintals of rice were distributed per month at a cost of Rs. 2 crores per month. In addition there was one time distribution of 4,698 quintals of sugar, 3,99,357 litres of edible oil, 30,08,883 litres of milk was also distributed till

May, 1985. It was decided to discontinue free ration in the municipal wards not directly affected by gas leakage.

So far Rs. 15,50,39,282 have been spent till October, 1985 on distribution of free rations. Another decision taken by the government was to give a consumption grant of Rs. 15,000 per family to those with family income of less than Rs. 500 per month. So far 18,000 families have been given the consumption grant of Rs. 1,500 each. It is estimated that about 48,000 families were qualified for this grant. A sum of Rs. 2.7 crores has been distributed as consumption grant to the affected people. An amount of Rs. 1.4 crores has also been disbursed as ex-gratia cash relief to the survivors of the gas tragedy. 2,222 cattle also died in the catastrophe. A sum of Rs. 2,28,478 has been distributed as compensation to the cattle owners.

Economic Rehabilitation

As a result of the disasterous leakage of toxic gas at Bhopal a large number of people were disabled and their physical capacities seriously impaired. The tragedy put out of gear the economic life of the people for a long time, affecting normal industrial, trading and business activities. The organised sector, specially the daily wage-earners suffered the most. The Bhopal Municipal Corporation is divided into 56 wards, of which 30 wards, with a population of 3,51,214 according to the 1981 census, have been identified as affected areas. Immediately after the catastrophe struck the town, a decision was taken by the government to conduct a Socio-economic survey of all the families living in these areas. Initially, a survey was conducted by the Tata Institute of Social Sciences of 25,000 families. A second survey was done by the Social Welfare Department of the government of Madhya Pradesh. The broad information which has been collected in the survey included number of deaths, number of persons disabled by the tragedy, adults who became non-active after the gas leakage, orphans, widows and families with no surviving members. The survey has now been completed and about 80,000 families have been included in it.

The survey will be useful for setting up numerous projects such as socio-economic and medical relief. The schemes for economic rehabilitation of the affected persons are being obtained. A large number of people have suffered serious impairment of their physical capabilities and can no longer work in professions involving hard physical labour. Most of the people in these areas belong to unorganised sectors with little or no skills. The economic rehabilitation plan proposes to include people in both the industries sector as well as service sector. The foremost need is to train people so that they can acquire new skills and be employed in works not involving hard physical labour. Training facilities have, therefore, to be provided immediately after which they can be helped to set up their own business. There will, however, always be people who may not be able to compete in the open market and such people will be provided employment directly through small public sector units or the units run by services organisations.

In the industrial sector, training-cum-development centres are being established which will provide necessary training for self-employment and also provide production facilities for those who can not set up their own business. For such people, who have set up their new production centres will serve as raw material banks and provide the outlet for their finished products. The production units to be established in the public sector by the government to generate employment should have an assured market so that these units do not run into problems of viability and sickness in future. There, proper marketing outlets had been ensured in the government or the public sector. It was, therefore, necessary to exercise a great deal of selectivity in the choice and location of the units both for training and production. The trades had been identified by keeping in view the fact that the people in these areas had very low skills and they needed employment which was not very exacting physically. Some of the trades which had been identified were tailoring, knitting, hosiery, carpet-weaving, handloom, dunnage manufacturing, TV repairs and assembly, soap making, agarbatti making, wax candle units, woollen blanket units, durry chalk manufacturing unit, stationery unit and carpentry unit. The other advantage of these industries was that they were pollution-free.

It was found that people in the gas affected areas, especially women could pick up the sewing trade quite easily. In order to provide marketing for their finished goods, the government of Madhya Pradesh gave the work of stitching of 3,50,000 school uniforms for children which were distributed by the government free of cost to the school children. In addition to this, the ministry of railways had also assured that the work of stitching of uniforms for the railways would be entrusted to these centres. A major garment factory with export orientation was also being set up in Bhopal. Negotiations were well underway with a leading export house for setting up this factory. Similarly, tie-ups had been made with other government departments, public sector undertakings and private sector industries.

The Electronics Corporation of Madhya Pradesh planned to set up a medium unit for assembling and manufacturing TV sets. Initially, the unit was proposed to be set up in the industrial estate of mandideep, about 20 km from Bhopal. This was now being shifted and persons from gas-affected areas who had high skills and aptitude for electronics would be trained in the unit as TV mechanics, etc. Every year training would be imparted to 100 persons. With their skill they would have an assured job potential.

People will also be helped for setting up their own business. There is a programme, special training and employment programme for the urban poor (STEPUP) in Madhya Pradesh. STEPUP has two broad features of self-employment through bankable schemes and training and execution of skills that would enable the beneficiaries to avail of the opportunities under the programme. STEPUP does not specify any minimum educational qualification for finances, nor is it designed for any highly sophisticated skills. Yet, it has wide range of functions such as carpentry, soap making, cycle repairing, motor rewinding, embroidery, tailoring, hosiery, welding, etc. which could be included in it, apart from petty shop keeping. As far as the gas victims were concerned, the choice of schemes would depend mainly on their mental and physical capability, apart from, of course, their aptitudes and traditional skills. Owing to the extraordinary situation in Bhopal, certain concessions had been made to enable the gas victims to take advantage of schemes under STEPUP. Eligibility for finance under STEPUP was subject to two conditions: the beneficiary should not be more than 40 years of age, and his family income from all sources should not exceed Rs. 4,500 per annum. In normal circumstances, average financing under STEPUP did not exceed Rs. 5,000 per case. Now, all these conditions had been relaxed for the gas victims. There was no age limit and the income limit had been revised to Rs. 5,000 per annum. Average financing had also been increased to Rs. 10,000 per case. There would be a component of 25 per cent of subsidy by the government on the amount of loan sanctioned by the bank. There was a proposal to take up the cases of 16,000 gas victims under the scheme.

Relief and rehabilitation

(up to 15 October 85)

Assistance to next of kin of dead		**Number**	**Amount**
Rs. 10,000	for each dead	1,400	Rs. 1.40 crore
Assistance of Rs. 1,500 per family to families with income under Rs. 550		12,000	Rs. 1.80 crore
Distribution of free foodgrain to slum-dwellers in the gas-affected areas (from 22-12-84 to 15-10-85)		—	
Wheat			488924 quintals
Rice			172779 quintals
Additional free supply of sugar edible oil and milk			
Sugar	4698 quintals		From 21-12-84 to 31-5-85 value approx. Rs. 64 lakh
Edible oil	399357 litres		From 8-12-84 to 31-5-85 value Rs. 1.08 crore
Milk	38,08,083 litres		
Distribution of nutritious food to expectant mothers and children			Through ?43 Anganwadis
Compensation for dead animals			For 2122 dead animals Rs. 22.84 lakh

Medical aid

Beds with modern electronic equipment at the M.G.M. Medical College	60 beds
Hospital in the gas affected area with facility for Pathology, X-ray etc.	30 Beds
Number of dispensaries in the gas affected area	17
Number of Ambulances in the gas affected area	9
Mobile dispensaries in the gas affected area	3
Door to door examinations of patients	50 doctors

Relief and assistance

Affected Population	3.20 lakh
No. of dead	1754
Patients admitted to Hospitals	1.73 lakh
Seriously afflicted	14,000

Future plans

Augmentation of medical facilities.

Six zonal claims with specialists for the 30 affected wards.

Special facilities for treatment of new born babies at the Kamla Nehru for women and medical college.

Expansion of 30 Bed Hospital to 100 Beds.

Establishment of school with Hostel for children

Augmenting the number of Anganwadis

Establishment of Five Industrial Training Centres in the affected area.

Establishment of Training-Cum-Production centres.

Loans and Grants for employment under the STEPUP Schemes.

Major scheme for upgradation of the Environment.

Administrative measures

Creation of a separate department of Gas Relief and Rehabilitation.

Wide administrative and financial powers to the department.

Posting of a large number of officers in the affected area.

Division of the affected area into seven administrative sectors.

Affected area constituted a separate medical district.

Offices for the Chief Medical Officer and the Additional Collector (Gas Relief) in the affected area.

About 50 to 80 work centres to train people and create production units were going to be set up. The people living in these areas tended to engage in non-formal sectors of employment. These training programmes would strengthen employment opportunities in this sector which would go a long way in strengthening the economy by a multiplier effect. Whilst it was not possible to upgrade the skills of adults in these areas to a very great extent, it was possible to train the children and young persons for medium and high-skilled industries. For this purpose, it was proposed to establish five industrial training institutes in the localities: two for girls and three for boys.

For the development, infrastructure had already been created in the form of bypass links, roads, etc. But no major work centre had been developed in the northern part of the city which was the most affected area. The closure of the Union Carbide factory created a void in the economic life in that part of the city in order to take advantage of the economic inputs in infrastructural development, particularly the distribution net work, it was necessary to develop major work centres. Thus, there was a proposal to develop commercial services and industrial sectors such as Transport Nagar, timber market, zonal market and clean services industries mainly servicing, repairs, electronics, etc. between Berasia road and Chhola road, development of industrial area on both sides of the Bhopal-Delhi railway line near the link road. This would help the affected population by giving them opportunities to become new entrepreneurs after the training in various skills.

MEDICARE OF BHOPAL GAS VICTIMS AN OVERVIEW

There was total chaos in Bhopal on the morning of December, 3, 1984. It was as if a thunderbolt from the blue had struck the city and left thousands of people helpless and bewildered. No one knew what had hit the city. Wave upon wave of human-beings continued to pour in the city hospitals till arrangements had to be made for people on the lawns, in the verandahs and every inch of available space was covered by patients writhing in agony. It was a formidable challenge to the medical profession and to the administration. In the absence of any toxicological literature about methyl isocyanate the doctors did not know what line of treatment to follow. There was complete silence from the Union Carbide management. They did not disclose anything about the toxic gases released from the plant and about the antidote. They had not disclosed anything even till todate.

The people who came to the hospital suffered mainly from breathlessness and choking, acute irritation of eyes and suffered gastrointestinal disorders. The doctors decided to give symptomatic treatment in these cases which had very beneficial effect. It was clear even in the early stages of clinical management of the patients that the health infrastructure should be adequately strengthened and expanded to meet not only the short term, more so the long term needs. Currently medical services to the patients exposed to the toxic gas were being provided by a string of dispensaries, special clinical examination-cum-treatment centres and hospitals. The Health department had set up a 30-bed hospital in the affected area with all necessary equipment and laboratory facilities. Medical specialists, gynaecologists, opthalmologists, surgeons, paediatricians, pathologists and radiologists had been stationed in this hospital.

Sixty beds were reserved in the medical college exclusively for the gas victims. There were 20 beds available in special eye hospital opened with the assistance of the Royal Commonwealth Society for the Blind. Special clinics had been set up at 17 places in the 30 affected municipal wards. Some of these were equipped with X-ray and laboratory facilities. The 30-bed hospital in JP Nagar, one of the worst affected areas, had been expanded by building a permanent shed with 12 rooms and spacious waiting room to serve as an out-patients' department. Indian Council of Medical Research (ICMR) was also running a laboratory in this hospital and equipment for blood-gas analysis, urine thiocyanate test and double blind study was available here. Among the dispensaries were five polyclinics where X-ray machines were available and specialists were posted for attending the patients.

Its Six dispensaries have well equipped laboratories. The medical College ward is equipped with the most modern facilities, some of which are not available anywhere else in the country. It is equipped with a computerised blood-gas analyser to study the capacity of lungs to oxiginate blood and discharge carbondioxide.

There are nine ambulances reserved exclusively for the gas-affected areas and in addition there are three mobile medical vans for these areas. The hospitals and dispensaries have been given sufficient stock of necessary medicines so that the people do not have to go to private doctors for treatment. Those who have suffered impairment of their vision have been given free spectacles.

The clinics and dispensaries have been provided with health cards for systematically recording the treatment given, in order to facilitate the follow-up action. Special documentation is being done for indoor patients and outdoor patients who were exposed to the gas and need long term treatment and follow-up. Special attention is being paid to pregnant women with regard to ante-natal care. The Indian Council of Medical Research has taken up a project for the follow-up of all pregnant mothers and the new-born in a population of 88,000 in 16 wards. It is proposed to follow up the growth—both mental and physical—of these children for a period of six months.

A psychiatric clinic had also been set up with the help of experts from Lucknow Medical College. More than 1,000 cases were so far registered here. The Indian Council of Medical

Research with the collaboration of the Medical College, had also taken up an epidemological survey of 17,000 families covering 85,000 persons in 13 affected wards. All these families were being followed up thoroughly by every fortnight visits made by the medical and para-medical staff. Those still found to be suffering from the after-effects of the toxic gas were provided treatment, at home or at the clinic or hospital as the case was. In addition to this, 20 research projects and studies had been taken up in collaboration with the Indian Council of Medical Research to assess the various possible effects of MIC on human beings.

Although detailed documentation about the ailments was available from the socio-economic survey as well as from the medical records maintained at the dispensaries, clinics and hospitals, it had been decided to carry out a fresh survey at this stage through house to house visits by medical teams. A team of 60 doctors had been deployed for the job. The doctors provided referral slips to such patients who were not being treated at the clinic and direct patients to report at the nearest clinic or hospital for proper treatment and follow-up action. Since the families taken up in the ICMR epidemological survey were being followed up by the Medical College teams, they were not covered by these doctors because it would lead to duplication.

In addition to these, several voluntary organisations came up for conducting such surveys. Recently a team of 10 doctors from Raipur came to Bhopal under the auspices of the Red Cross Society, Raipur branch, and Akhil Bhartiya Rachanatmak Samaj, Raipur. There were several members of legislative assembly in Bhopal who were qualified medical doctors. For such doctors who were members of the Legislative assembly also went round the affected area and assessed the health status of the patients. The symptoms of tuberculosis among some patients were aggravated after exposure to the gas. Special efforts were being made to detect such cases and provided proper treatment to them.

The areas, which were affected, were mostly inhabited by poor people. The exposure to the gas weakened them. The government therefore started a mass immunisation programme for children in this area. Apart from medical treatment, it was also proposed to start shortly two Yoga centres in this area for treatment of lungs by teaching breathing exercises. A well equipped clinic had been established for long term follow-up of patients to detect any possibility of after-effect of MIC on eyes. The clinic was equipped with modern instruments such as Corneal Specular Microscope, Slit Lamps, Fundus Camera and Opthalmoscopes for the study of the anterior and posterior parts of the eye. ICMR sponsored and financed a study for long term follow up of the occular changes in gas affected cases. Instruments such as electro-retinogram and electro-occulogram were being procured for further studies.

More than 3,000 patients were reporting every day at every clinic and dispensary. It was clear that the health facilities available in the affected areas would not be able to provide all the medical support necessary for the affected population. Steps had therefore, been taken to strengthen the medical infrastructure. A new 100-bed hospital, at a cost of Rs. 2.5 crores was already under construction. This would be equipped with the most modern and sophisticated equipment. The capacity in the medical college was also being expanded. Modern facilities were being introduced in the Kamala Nehru Women's Hospital for the care of mothers and infants.

Sub-acute syndromes of MIC toxicity had now become evident among those who had been most exposed but had survived. Although the main symptoms still related to the respiratory tract, symptoms involving other organs were also being observed. The different symptoms could be grouped into the following main areas:

Medical Research Projects

Project Code No.	Title of the Project	Principal Investigator
5/9 2/85-NCD-1	Long term studies on the health effects of MIC exposure through community health clinics.	Prof. S. N. Sharma

5/9 3/85-NCD-1	Clinical studies including respiratory function tests among MIC-affected adults.	Prop. N. P. Mishra
5/9 4/85-NCD-1	Follow-up study of children (5-15 yrs.) affected by MIC exposure.	Prof. N.R. Bhandari
5/9 5/85-NCD-1	Follow-up study of occular changes in MIC-affected subjects.	Prof. P. C. Mittal
5/9 6/85-NCD-1	Clinico-pathological and histopathological investigations of MIC-affected persons.	Prof. B. S. Darbari
5/9 7/85-NCD-1	Microbiological investigations into the pulmonary complications of the MIC-affected subjects.	Prof. Darbari and Prof. K.A.V.R. Krishnamachari (Hyderabad)
5/9 8/85-NCD-1	Clinical and forensic toxicological studies in MIC-affected persons.	Prof. Heeresh Chandra
5/9 10/85-NCD-1	Radiological manifestation in acute MIC gas episode in chest.	Prof. P. N. Sharma
5/9/16/85-NCD-1	ICMR MP Government establishment of data base information system for MIC follow-up studies at Bhopal.	Prof. S. N. Sharma and Dr. K. Ramchandran
5/9/16/85-NCD-1	Genetic risk evaluation of MIC cytogenetic studies in population exposed at Bhopal.	Prof. R. K. Jain
5/9/23/85-NCD-1	Immunological studies of MIC.	Prof. B. S. Darbari

5/7/28/84-RB	Teratogenic and mutagenic effects of MIC on humans.	Prof. N. R. Bhandari
5/9/17/85-NCD	Metabolic studies in MIC-affected population	Prof. S. A. Bhambhal (Bio-chemistry)

ICMR Projects Sanctioned for Institutions other than Gandhi Medical College, Bhopal

Sequential study of pulmonary and neurological changes in exposed subjects.	Dr. S. R. Kamath Bombay
Cytafluorimetric studies of cells in MIC exposed subjects.	Dr. M. G. Deo Bombay
Study of pulmonary function tests including blood gas analysis.	Dr. P. S. Narayanan
Neurological manifestations of MIC poisoning.	Dr. E. P. Bharucha Bombay
Mental health studies in MIC-affected population.	Sr. B. B. Sethi

The Arizona State University (ASU), which helps in running the Bhopal technical and vocational training centre (BTVTC,) has pulled out of Bhopal. Stating this, Mrs. Indira Iyenger, managing director of the non-profit making BTVTC, told that the ASU took this decision in view of continued hostile and non-cooperative attitude of both the Union and state governments towards the centre. She was informed about the decision of the ASU authorities. Mrs. Iyenger said that the Union home minister had put a ban on the centre accepting any foreign funds without prior permission and the external affairs ministry had refused to renew the visas of American professors visiting Bhopal in connection with the centre's activities. She made repeated representations to both the ministries but without success.

This decision, she said would adversely affecct about 300 students who were learning various crafts to augment the meagre income of their families. The centre had already purchased a number of equipment. Mrs. Iyenger admitted that the centre, which was formally inaugurated on June 16, 1985 was under a cloud of suspicion and mistrust. This was because the Union Carbide Corporation (UCC) had advanced $2 million (Rs. 2.5 crores) to the ASU to support its plans to set up the vocational and technical training centre in Bhopal. Even the UCC chairman, Mr. Anderson, had made it clear that such financial support to the ASU would not prejudice the compensation case.

She said that it was unfortunate that the government first gave the ASU the green signal, and once the centre started functioning the former developed an attitude of hostility and non-cooperation. She would rather welcome the government straightway telling the ASU to close down the centre rather than "indulge in bureaucratic pinpricks."

Intelligence officers had visited the centre regularly and the municipal corporation had demolished some new construction. Mrs. Iyenger who is also president of the MP Christian Association, denied that there was anything "anti-national or unpatriotic" in accepting money from the multinational. The centre had no direct link with the UCC. She pointed out that besides BTVTC, the MP Red Cross Society, of which the governor is president, had accepted Rs. 6.5 crores from the UCC via the American Red Cross. The Prime Minister's relief Fund had directly accepted $1 million (about Rs. 1.2 crores) from the UCC soon after the gas tragedy. If the UCC money was tainted, would they have accepted it? She asked.

Mrs. Iyenger said that the centre had so far received only Rs. 16 lakhs from the ASU (before the government ban) of which they had spent Rs. 8 lakhs on establishment. She pointed out that the centre was not meant exclusively for gas victims. Its aim was to provide training to the local people in various crafts. The centre located in a farm overlooking the lake, had already paid Rs. 1 lakh for a one-year lease to the owner, Mr. Baba Walia. Mrs. Iyenger said that it would be difficult to run the centre after the withdrawal of ASU support, but she would make determined efforts to see that at least the first batch of trainees stood on their legs. Currently, that was her primary concern.

CONCLUSION

A CONTINUING TRAGEDY

Doctors of the Indian Council of Medical Research (ICMR) have expressed concern at the gas victims of Bhopal gradually losing their eye sight. This is one of the deadly effects of the methyl isocyanate (MIC) gas that leaked from the Union Carbide plant killing at least 2,500 persons and affecting four lakh people. Dr. Krishnamurthy, chairman of the high-powered Scientific Commission, appointed by the Union government in 1985 to go into the effects of MIC on human beings, flora and fauna, air and water, told that the loss of vision was appearing at a very late stage, this was both disturbing and serious. He stressed the need for a more intensive study of the problem.

It was stated that a majority of the gas victims had corneal opacity (blurred vision). The doctors stressed the need for a vigorous eye-check programme on a mass-scale to prevent any further damage to the cornea. The commission is understood to have expressed its dissatisfaction at the state government giving priority to the compensation issue at the cost of the medical teratment of the gas victims. This observation came when the ICMR doctors said that the victims continued to suffer from respiratory ailments and general weakness.

The Madhya Pradesh government's approach to the Bhopal disaster by way of suppression of information, inadequate medical relief and persecution of voluntary relief workers came in for severe criticism both in the press and the public:

> "In the traumatised conditions in Bhopal, the mind of bureaucracy and the police have become so closed that they can not conceive, let alone concede or listen to non-governmental people working for the gas victims," said Prof, J.P.S. Oberoi of Delhi University's sociology department.

who had recently been to Bhopal as a member of a citizen's Fact Finding team after the arrest of three voluntary workers on 'spy charges.'

Dr. Vijay Verma, of Delhi University's physics department, said that the recent arrest of voluntary workers in Bhopal on "trumped up charges" and the harassment of many others, including doctors, exemplified the state government's attempts to suppress information. The Motilal Vora government should hang its head in shame over the manner in which it has allowed the state machinery to be misused for wrong purposes. It has willingly allowed some local doctors, bureaucrats and police officers to harass local relief workers who have been doing creditable work among the Bhopal gas victims.

The state government's own incompetence in dealing with the gas victims and their families is therefor everyone to see. Now by arresting two or three activists from voluntary groups which have been conducting effective relief work with the gas victims and orchestrating a spy hysteria campaign against them, it has only managed to erode its own credibility further. According to some reliable sources the motivation for these arrests and the spy-drama seems to be two-fold:

First, to cover up the failure of the state government doctors even to keep proper medical records of the gas victims. Indeed, this was the subject of discussion at the meeting at which senior Bhopal doctors manhandled the activists of the voluntary organisations who were taping their discussions. This meeting incidentally was open to non-governmental personnel and a number of private doctors attended it.

Secondly, the aim seems to be to prevent the disputes that exist between two groups of government doctors from becoming public. The petty differences in fact, have marred the evolution of a proper line of treatment and monitoring of the gas victims right from the beginning.

RED CROSS CAUGHT ON WRONG FOOT

The Indian Red Cross Society is caught in a fierce controversy over the way it has spent the money it received from the Union Carbide Corporation (UCC) through the American Red Cross. The M.P. Red Cross (MPRC) which has so far received only Rs. 52 lakhs, has bought medical equipment whereas the entire amount was meant to be distributed as monetary relief to the gas victims of Bhopal.

The Union Carbide had made available on November 27, 1985 a sum of 5 million or about Rs. 6.4 crores at the behest of Mr. John F. Keenan, judge of the federal district court for the southern district of New York, to provide interim monetary relief to the gas victims.

Since the government of India refused to accept any responsibility to account for this amount, the American Red Cross undertook the responsibility for the distribution of this relief to the Indian Red Cross through the International Red Cross. Todate, $ 2 million or about Rs. 2.5 crores has been transferred to the Indian Red Cross. But there is no record of how the money has been utilised and for what purpose. Of the Rs. 2.5 crores which the Indian Red Cross has received, it has passed on merely Rs. 52 lakhs to the MPRC. The state unit, instead of pending the money on providing cash relief, has spent it on buying costly medical equipment, vehicles and overheads. It has not provided monetary relief to the gas victims as directed by the judge at the time of giving the amount. Part of the money has been spent on running three dispensaries and a clinical investigation centre.

However, Dr. S.K. Vashishtha, project director for the utilisation of this amount, denied, having any knowledge about the court directive for using the money for providing monetary relief to the gas victims. He even refused to accept that the funds he was spending had come from Union Carbide. He said that so far as the MP Red Cross was concerned, the money was given by the Indian Red Cross and it had not put any condition whatsoever about providing monetary relief to the gas victims. Following the footsteps of the parent body, the

MPRC was caught in the fierce controversy over the floating of a lottery and the governor, Prof. K.M. Chandy, who is its president, had dissociated himself with the body. Only recently, the governor revived the MPRC when its controversial chairman, Mr. R.P. Nagu (whose firm had a contract of providing security guards to Union Carbide's Bhopal plant) was replaced by Brig. K.P. Pandey.

It is strange that neither the governor, as president of MPRC, nor any other office-bearer is anxious to inquire from the parent body—the IRC—about what happened to the remaining Rs. 2 crores meant for the Bhopal gas victims. "What do we have to believe except press reports, that the IRC has Rs. 2.5 crores", one senior official of MPRC said and added that they have no official information in this regard. According to one report, the IRC has invested the amount in fixed deposits with different banks without the approval of the concerned authorities and is in no hurry to hand over the amount to MPRC.

It is also pointed out that in the aftermath of the gas disaster, the IRC as well as the MPRC had received numerous donations. So far no detailed account of the donations received has been published. The IRC is understood to have suggested to the MPRC to adjust the donations against the funds received from the Union Carbide for the Bhopal gas victims. According to the 20-page "project outline in support of the gas victims of Bhopal" marked "confidential" and prepared by the IRC, it is proposed to set up a clinical investigation and pulmonary physiotherapy centre for assessment and treatment of pulmonary functions structure of damage to lungs at a cost of Rs. 1.53 crores. The centre will have, among other things, radiology, pathology and physiotherapy equipment costing Rs. 1.4 crores. A sum of Rs. 9 lakhs will be spent on accommodation and Rs. 4.45 lakhs on vehicles and furniture. The recurring expenditure would be Rs. 35 lakhs per annum.

The other expenses shown in this project are Rs. 34.80 lakhs on static medical units, Rs. 26.15 lakhs as recurring expenditure, Rs. 10.80 lakhs on mobile medical units, which includes Rs. 2 lakhs on medicines, Rs. 8.80 lakhs on vehicles apart from over Rs. 15 lakhs as recurring expenditure per annum. Although the project director, Dr. Vashishtha, denies having any directive about providing monetary relief to the gas victims, the project report has a provision of Rs. 90 lakhs to be provided as monetary relief to 5,000 families of gas victims at the rate of Rs. 150 per month per family.

However, the Bhopal tragedy continues and thousands of people are afflicted by lung, eye and skin diseases besides psychological problems. The official agencies categorised the gas victims as severely affected (65,000), mildly affected (20,000) and those under controlled effects (14,000). The abortion, mortality and infant mortality rates in the severely affected category have been found to be three to four times higher than in the controlled category during April, 1985 to March 1986. While official agencies' reports, kept confidential, point to a very serious situation, the government wants people to believe that all is well in Bhopal. In fact, statistics bleed with reality.

According to the Indian Council of Medical Research (ICMR)'s report on Gynaecological disorders, out of the 86,153 worst affected population, 2,698 women were pregnant as on December 2, 1984, of the 2210 births, 28 babies had congenital malfunctions, 150 babies died at birth and there were 402 abortions. Within the first week of their lives, 78 babies died and 36 infants had only a life span of less than a fortnight. Within a month of their births, one or more in four babies born in the gas affected areas were either aborted, still born or dead. So the tragedy still continues. . .

Investigations show that there has been a significant number of abortion after December, many more than the figure claimed by the government. The state health minister's statement that only 36 women had aborted between December 3 and June 15 could be taken with a pinch of salt. Since Januray 1985, there have been over 1,000 foetus or babies cremated, representing a considerable hike on the previous year's figures.

According to crematorium officials till July, 1985, four deaths of aborted or still born babies took place daily. Obviously, a large number of these deaths was due to gas exposure. In Bhopal today, there has been virtually no access to medical information from the hospitals or official agencies and the government statistics are under-estimated. Nevertheless, a sufficiently high number of women from the affected areas suffer from acute pelvis infalmatory disease (P.I.D) leucorrhoea (white vaginal discharge) and excessive and irregular menstrual bleeding.

A noted Gynaecologists Dr. Rani Bang showed from a small sample study conducted in 1985 that over 80 per cent of the women examined suffered from P.I.D.; while normally the usual proportion for women in the same socio-economic strata would be approximately 30 per cent. Complications leading from P.I.D. can lead to ectopic (or tubal) pregnancy and also infertility. Cases of cervical erosion and the backward bending of the uterine had also been found. For many women, with repitorial problems coupled with excessive menstruation had led to a feeling of inertia, fatigue and ill-health. The report that was published at the end of these findings suggests a strong enough reason to look into of these medical complaints. Kamala Devi resident of J.P. Nagar just sits in the sun these days. She can not look after her own children because of excessive discharge and the medication recommended at the government hospital is not really helping. After strenous long waits in queues she has now decided to go in for private treatment as she feels, that the *sarkari* doctors don't take her condition seriously. This is not the complaint of only woman. Many of them in fact are forced to go in for private investigations, because medical facilities in government hospitals are just not adequate enough to tend to their needs. Consequently, these victims are not recorded in the official data and their needs are ignored completely.

Indeed the Jawahar Lal Nehru Hospital, situated near the affected areas, was opened specifically last year to cater to the needs of those hit by the MIC gas. However, most of the gas affected victims end up travelling a considerable distance to Gandhi Medical College or the Sultania Zenana hospital. For many, it is not an easy existence:

> "Every month, I spend Rs. 300-400 on treatment alone", says Durga Bai who is employed as a maid servant," I have

decided to go in for private treatment, because if I go to government hospitals, I'll be there till 2 o'clock eyeryday. As it is my husband can't work and I'll also lose my job if I am absent daily". For most it's a case of finding employment, which they are able to do. Ironically, those people wanting to get back to their original position are turned away with a callous reply: "You can't do this job any more, you've been affected by the MIC gas."

And if the tragedy of the mothers is not enough, the Carbide child has also a dismal future. Every second child effected by MIC complaints of breathlessness and a feeling of fatigue, at the least exertion. One out of every five suffers from a persistent, hacking cough, which not even medical treatment can bring it to normal. Almost all suffer from a lack of appetite which is hampering their growth. One out of every four conceived 9 months after the gas leak, suffers from blisters, multiple boils and blurred vision. One child in every six has developed purple patches on the face. A 18 month child was born with skin discolouration—a common enough complaint.

Deformities are another major horror to which the babies of Bhopal are subjected to. Deformities include not only the government's definition, that is external deformities but internal injuries, such as haemorrhoages in the lungs and brain—which are fatal. Saraswati in Kanchi Chhola, delivered a child in April, 1985, publicised as a freak child, the baby, after its eye lids were unglued literally had red spots instead of eyes. The cause of death, as diagnosed by Professor Hareesh Chandra, Head of the Department of Forensics at Gandhi Medical College, was due to poisoning. Saraswati's baby is not an exception which proves the rule.

There has been an increased rise in the case of deformed babies. Eventhough Dr. I.G. Bhandari, head of the Paediatrics Department and chief project co-ordinator of ICMR said:

"The 30 cases of congenital malformation in the babies born to 2,504 mothers, is well within the average population ratio." But despite government postulations to the contrary, the birth of deformed babies is ceratinly not normal. Private clinics in Bhopal ascertain and have documanetary proof of numerous still born babies that have died because they have been exposed to blood carrying less oxygen and blood poisoning, that is hypoxia and histo toxity. Similar accounts are narrated by mid-wives and *dais*. It is these children who have been ravaged by physical disabilities and who do not come under the '648' official dead MIC children. Already weakned by pulmonary afflictions, their immunity aganist disease is low. Tuberculosis and jaundice have already spread in the shanty dwellings. Inspite of Dr. Bhandari's claim that such children are receiving VVIP treatment, the children, especially the babies are very listless in the official accounts.

Many have not begun to walk, though they have completed their first year. And moreover the effect of lack of oxygen to the baby's brain during the gestation period will be seen only later. Commented a doctor, who wanted to remain unidentified:

> "There is a strong possibility of retardation in some children. But only time will tell."

Among most people, however, there is an astounding presence of anger directed against Union Carbide. Some who are alive may be happy to have escaped death. While others have vague ideas of an el dorado, conjured up by vision of compensation. But however much the compensation, no money can compensate for the sorrow or the ill-health caused by the US multinational Union Carbide to the Bhopal mothers and their unfortunate children due to the hazardous gas-walk on their lives. Silently they will bear their sufferings upto the days of the fading glow of our posterity, for no fault of their own. History will remain the mute witness to this avoidable disaster and an unending human suffering.

SELECT BIBLIOGRAPHY

2-3 December in Bhopal, *Massline*, 11 (4-5), January-February, 1985:1.

100 days after Bhopal, *Energy Futures* 1 (4), April, 1985:5, Reports from Nairobi Conference.

375 killed by escaping Gas, *Times*, December 4, 1984.

20,000 Gassed to Death, *Massline*, 11 (4-5), January-February, 1985: 1,4.

Abraham, Amrita, Third World Night Moves, *Sunday Observer*, December 16, 1984:8.

Abraham, Jugu: When Air Became Poison, *Weekend Review*, 5 (41), December 16-22, 1984:3.

All about UCIL, MIC and the Night of December 2, *Nagpur Times*, December 21, 1984.

Alvares, Claude: Bhopal tragedy: Climax of a Hazardous System, *Asian Pacific Environment*, 3 (1), March, 1985: 1,7.

Alvares, Claude: Fate of the earth, *Illustrated Weekly of India*, 105 (34) December 30, 1984: 14-17.

Alvares, Claude: Sage of a Hazardous System, *Afkar Inquiry*, 2(2), February, 1985: 18-20.

Alvares, Claude: Union Carbide and Indian Government, *Asian Pacific Environment*, 3(1). March, 1985:7-8, *See also* pp. 9, for Dr. B. E.K. Bal's article.

Amrit Bazar Patrika, Death in the Air, *Amrit Bazar Patrika*, December 5, 1984: *Editorial*.

Antony, M.J.: Ignorance Exploited to Evade Compensation, *Indian Express*, December 6, 1984.

Any which way out of Bhopal, *New York Times*, December, 16, 1984.

Arjun Singh: *Press interview*, Bhopal, December 6, 1984; *Patriot*, December 7, 1984.

Asia-Pacific People's Environment Network: Appeal to People, *APPEN*, March 1985.

Bakshi, P.M.: Bhopal tragedy and Indian courts, *Statesman*, April 3, 1985.

Banerjee B.N.: Bhopal Gas tragedy—Accident or Experiment, Paribus Publishers and Distributors, New Delhi, 1986.

Bandyopadhyaya, Jayanta: Technological terrorism, *Indian Express*, December 25, 1984.

Belli, Melvin M.: Bhopal tragedy and US lawyers, *Indian Express*, January 10, 1985; January 11, 1985.

Benedict, K.: Accident or subotage, *Weekend Review*, 5(42), December 23-29, 1984:3.

Benerji, Debabar: Epidemiological and sociological study of the Bhopal tragedy, *Medico Friend Circle Bulletin*, January 1985:3, J.N.U. Study. For more studies, *see also* pp. 5,8.

Bergin, Edward J.: Interview on Bhopal Gas Tragedy, *India Today*, 10(2), January 31, 1985:102.

Bhagat, Dhiren: Only MIC leaked, *Spectator*, 253 (8162), December 15, 1984:8.9.

Bhagat, Dhiren A: Night in hell, *Sunday Observer*, December 16, 1984:20.

Bhandari, Arvind: Union Carbide Compensation, *Tribune*, January 16, 1985.

Bhed, K.N.: Compensating Bhopal Victims, *Indian Express*, January, 1985.

Bhattacharya, Biswanath: US Government intervened in a big way, *Amrita Bazar Patrika*, December 11, 1984.

Bhopal: Balance Sheet of Death, *Guardian Weekly*, 131(25), December 16, 1984:1.

Bhopal: Genesis of Massacre, *Massline*, 11(4-5), January February, 1985:3, 10. *See also* pp. 12.

Bhopal: Industrial Genocide? Kowloon, Hong Kong, Hung Hom Bay Centre, 1985. (Collection of articles and Newspapers.)

Bhopal: Log Keede Makodon Ki Tarah Mare Gaye, *Dinman* 20(51), December 16-22, 1984:8—12. *See also* pp. 13-17. For more articles.

Bhopal disaster aftermath, *Medico Friends circle Bulletin*, June, 1985: 2.3. Survey of disaster Area.

Bhopal imping Back to normal, *Amrit Bazar Patrika*, December 7, 1984.

Bhopal Main Jahrili Gas Ka Kahar, Kya Yah Shadyantra Tha, *Preksha* (Hindi) 1(8), December 16-23, 1984:1,9,14.

Bhopal Shadow on vital farm input, *Hindustan Times*, April 4, 1985.

Bidwai, Praful: Deadly delay in Bhopal, *Times of India*, December 19, 1984.

Bidwai, Praful: Failure of systems: Lapses led to mishap, *Times of India*, December 8, 1984.

Bidwai, Praful: Health damage alarming wide spread, *Times of India*, March 25, 1985.

Bidwai, Praful: MIC Proving deadlier than feared, *Times of India*, January 6, 1985.

Bidwai, Praful: Plant design badly flowed, *Times of India*, December 27, 1984.

Bidwai, Praful: Plant undermanned run down, *Times of India*, January 1, 1985.

Bidwai, Praful: Poisoned city, *Times of India*, December 15, 1984.

Bidwai, Praful: What caused the Pressure Buildup? *Times of India*, December 26, 1984.

Bilder, Richard B.: Lumpsum settlement for Bhopal, *Sunday Observer*, January 26, 1985.

Bose, Ajoy: Blindness feared for survivals, *Guardian Weekly*, 131(25), December 16, 1984:6.

Brident, John.: U.S. lawyers and Bhopal disaster. *Times of India*, March 2, 1985.

Brummer, Alex: Third rate standard for Third World?, *Guardian Weekly*, 131(25), December 16, 1984:7.

Butler, Sharon: Enter the ambulance clasers, *Sunday* 12(8), December 23-29, 1984:62.

Campaign Alert to Environmental NGo's world-wide: Fight toxic waste problems, *APPEN*, May, 1985: 1-7.

Canada, Air Pollution control directorate, Environmental protection service, Canada's Air Pollution control Programme:

Report EPS, Ottawa, Information Canada, 1975, 46 pp.

Carbide Bluffed Authorities, *Hindustan Times*, December 7, 1984.

Carbide chief told to leave India, *Hindustan Times*, December 8, 1984.

Carbide extends relief, *Economic Times*, December 4, 1984:1,6.

Centre for Science and Environment: Appeal for help, New Delhi, April 14, 1985.

Chakrapani, R.: Aides contradict Union carbide chief's claim, *Hindu*, December 13, 1984.

Chakrapani, R.: House panel for uniform safeguards, *Hindu*, December 14, 1984.

Chakrapani, R.: US Plans having fresh look at safety measures, *Hindu*, December 6, 1984.

Chakravarty, Pratap.: Workers kept in dark about killer gas, *Partiot*, December 14, 1984.

Chandra, Feroze R.: Breathing death in Bhopal, *Herald Review* 1(14-15), December 9-22, 1984:8-11, 14.

Chattopadhyaya, Boudha Yan: Bhopal, the face of the Indian accomplice, *Amrita Bazar Patrika*, December 21, 1984.

City of Death, *India Today*, 9(24), December 31, 1984:4-25. Articles and Reports by M.O. Srikant Khandekar, Suman Dubey, Indenit Budhawa, Madhu Trehan.

Claiborne, William: U.S. Executives confers with Indian official, *Washington Post*, December 9, 1984.

Commerce Carnage by Union Carbide, *Commerce*, 149(3837), December, 15, 1984:909, *Editorial*.

D'Monte Darryl: Cloud of death, *Sunday Observer*, December 9, 1984:12.

Dave, Rohit: Environment and Economic Growth, Bombay, Commerce, 1973. 30 p.

Davis, Ian. Squatters: Who live next door to disaster, *Guardian Weekly*, 131 (25), December 16, 1984:7.

Dayal, John: Stones of MIC neutralised, *Patriot*, December 17, 1984.

Dayal, John: Accident or chemical warfare, *Link*, 27(19), December 16, 1984: 8-9, 11. *See also* pp. 10.

Dayal, John: People welcome Anderson's arrest, *Patriot*, December 10, 1984.

Delhi committee on Bhopal Gas tragedy: Repression and apathy in Bhopal Delhi, *Delhi Committee on Bhopal gas tragedy*, 1985.

Deshpande, Anjali: No gas is a killer: It is men who kill, *Probe India*, 6(11), January 1985: 31-33.

Deva Kumar, C., Killer Gas, *Hindustan Times*, December 7, 1984.

Deva Kumar, C. and Mukherjee, S.K.: Methyl isocyanate, Profile of a killer gas, *Science Today Journal*, January 1985: 10-11, 16.

Dewan, Manorama: Multinationals or mass Murders? *Democratic World*, 15(52), December 23, 84:5. *See also*, 6-7.

Diamond. Stuant: Bhopal tragedy: Operation design of plants are faulted, *International Herald Tribune*.

Doing Business in Third World: Chemical leak prompts debate, *New York Times*, December 16, 1984.

Dudani, A.T.: Dangers of Agrochemical residue, *Amrita Bazar Patrika*, December 28, 1984.

Dutt, Vijay: Sabotage or Accident? *Probe India*, 6(11), January 1985: 34-35.

Economist: Gassed in Bhopal, *Economist*, 293(7372), December 15, 84:14 *Editorial*.

Exodus from Bhopal continues unabated, *Amrita Bazar Patrika*, December 15, 1984.

Experts views on Gas effects disputed, *Times of India*, January 7, 1985.

Exporting Hazards, *India Today* 10 (2), January 31, 1985: 100-04.

Fera, Ivan: Another Environmental Threat, *Illustrated Weekly of India*, 105 (38), January 27-February 2, 1985: 50-51.

Fera, Ivan: The day after, *Illustrated Weekly of India*, 105 (34) December 30, 1984: 10-13.

Fernandes, Allwyn: Need for Poison information centres, *Times of India*, December 16, 1984.

Fishlock, Trevor: This has been our Hiroshima, *Times*, December 10, 1984.

Gandhi Peace Foundation—Report of the Discussion on Bhopal Tragedy. The responsibility of Scientist Community, New Delhi, May 14, 1985: pp. 5.

Gas leak Kills 335 in India, *Washington Post*, December 4, 1984.

Gas Poisoning toll amounts to 1600, *Times of India*, December 6, 1984.

Gas Tragedy toll exceeds 2000, *Times of India*, December 7, 1984.

GATT: Pollution control and international Trade.

Gaya, Javed: Bhopal legal implications, *Business India*, (77), December 17-30, 1984: 67-76.

Gaurishankar, V. : Course of Justice in U.S., *Hindustan Times*, February 19 1985.

Gee, David, etc. : Bhopal: 100 days on, *Manchester Guardian Weekly*, March 24, 1985 : 2.

Gemini, *Psaud* When death was in the air, *Sunday Midday* (234), December 16, 1984 : 10.

Ghosh, Sailen: Neglected Issues, *Hindustan Times*, January 18, 1985.

Gopal Krishnan, K. and Kavi, Ashok Row.: Gassed, *Week* 2 (52), December 16-22, 1984 : 14-20. *See also* p. 25-27.

Gopal Krishnan, K: France that followed, *Week* 3 (1), December 23-29, 1984 : 12-13. *See also* pp. 14.

Goswami, Nirmal: Bhopal Gas chamber, *Sunday Observer*, December, 16, 1984: 20.

Govt. that kills, *Organiser* 36 (31), December 16, 1984: 1,16.

Goyal, K.N.: Bhopal holocaust and the law, *Indian Express*, December 24, 1984.

Goyal, K.H.: Justice after Bhopal, *Statesman*, December 23, 1984; December 24, 1984.

Goyal, K.N.: Modalities for compensation, *Hindustan Times*, December 17, 1984.

Gupta, Joydeep: Question of Phosgene, *Stateman*, December 20, 1984.

Gupta, Y.P.: Poison armed and within, *Statesman*, December 15, 1984.

Hazarika, Sanjoy: On the eve of disposal Indian city lies silent, *New York Times*, December 16, 1984.

Herald Review: Bhopal: will it be last? *Herald Review*, 1 (14-15): December 9-22, 1984:6, *Editorial.*

Hindu: Bhopal Tragedy, *Hindu*, December 5, 1984, *Editorial.*

Hindu: Legal aid to Bhopal gas victims, *Hindu*, January 4, 1985.

Hindustan Times: Adequate compansation, *Hindustan Times*, December 6, 1984, *Editorial.*

Hindustan Times: Bhopal coverup, *Hindustan Times*, March 28, 1985.

Hindustan Times: Bhopal options, *Hindustan Times*, March 20, 1985.

Bhopal Melodrama: *Hindustan Times*, December 15, 1984, *Editorial.*

Development without dire, *Hindustan Times*, December 9, 1984, *Editorial.*

Grim tragedy: *Hindustan Times*, December 5, 1984, *Editorial.*

Now to be done, *Hindustan Times*, December 19, 1984, *Editorial.*

Temper of Science, *Hindustan Times*, January 10, 1985, *Editorial.*

After disaster, *Amrita Bazar Patrika*, December 10, 1984, *Editorial.*

Home truths about Bhopal, *New Scientist*, December 13, 1984:2.

India Centre for Science and Environment: State of India's Environment 1984-85, Delhi. Author deals with Bhopal Gag Tragedy on pp. 205-222.

India soliditary Group: Paris Appeal for Bhopal, Paris, January 16, 1985.

Indian Express: Bhopal suit, *Indian Express*, February 22, 1985, *Editorial.*

Indian Express: Deplorable Blunder, *Indian Express*, December 10, 1984, *Editorial.*

Indian Express: Perilous aftermath, *Indian Express*, December 4, 1984, *Editorial.*

Indian Express: They still need succour, *Indian Express*, December 24, 1984, *Editorial.*

Indian Express: Politicking Bhopal?, *Indian Express*, January 2, 1985, *Editorial*.

Indian Express: Treating Bhopal Victims, *Indian Express*, January 31, 1985, *Editorial*.

Indian Nation: By Tragedy, *Indian Nation*, December 5, 1984, *Editorial*.

Industrial Air Pollution, *Factory*, 123 (10), 1965.

Iyer, Pico.: Clouds of uncertainty, *Time*, 24 (26), December 24, 1984: 16-17.

Iyer, Pico.: India's night of death, *Time*, 124 (25), December 17, 1984: 8-13. *See also* pp. 5, 14-19.

Jain, N.C.: World poets speak up for Bhopal, *Statesman*, March 18, 1985:5.

Jethmalani, Ram: Guilty in Bhopal, *Indian Express*, April 19, 1985.

Jha, Prem Shankar.: Bhopal Tragedy, *Times of India*, December 7, 1984.

Keswani, Rajkumar: Bhopal's killer plant, *Indian Express*, December 9, 1984.

Khandekar, Sreekant: Grey Areas, *India Today*, 10(1), January 1-15, 1985: 101.

Khandekar, Sreekant: Legal damage, *India Today*, 10(1), January 1-15, 1985: 100-4.

Khandekar, Sreekant: Painful indecision, *India Today*, 10(2), January 31, 1985: 84-85.

Killer multinational vs killer congress I Govt, *Organiser*, 36(32), December 23, 1984: 1,4. Reprinted from *Patriot*.

Kottary, Sailesh: Whose life is it anyway ? *Illustrated Weekly of India*, 105 (34), December 30, 1984: 6-9.

Krishna Iyer, V.R.: Bhopalshima and Indian law, *Indian Express*, April 18, 1985.

Krishnan G.V.: Manual lists lethal chemicals in MIC Unit, *Times of India*, February 6, 1985.

Krishnan, G.V.: Murky Politics in Bhopal Stir, *Times of India*, January 16, 1985.

Krulilla, John W.: Some Environmental effects of Economic development, *Deadelus*, Fall 1967.

Kulkarni, Sudherendra: In Bhopal's long shadow, *Science Age*, 3(1), January 1985: 31-34.

Kusumakar, H.: Dire dangers of pesticide use, *Times of India*, December 8, 1984.

Legal Quagmire: Sought by all parties, Settlement on Bhopal seems more elusive, *Wall Street Journal*, April 12, 1985.

MacKenzie, Debora: Chemistry behind Bhopal's tragedy, *New Scientist*, December 13, 1984: 3-4.

Mahajan, Krishan: Union Carbide: Lawyers disgrace, *Hindustan Times*, February 11, 1985.

Martin, Patrick and Milner Brian: Safety lax at Indian Plant: Union Carbide chief says, *Globe and Mail*, May 1, 1985: 14.

Martin, Patrick: Future of Bhopal claimants in hands of New York Judge, *Globe and Mail*, April 16, 1985:13.

Medico Friend Circle Bulletin: Challenge of Bhopal, *Medico Friend Circle Bulletin*, June 1985: 1, 8. *Editorial*.

Mehta, Balraj: Might of the multinational, *Tribune*, December 18, 1984.

Mehta Vinod: Lessons of Bhopal, *Nagpur Times*, December 21, 1984.

Menon, Kesava: Calamity beyond comparisons, *Frontline* 1(2), December 15-28, 1984: 59-65.

Mitra, Manojit: Shutters down all over Bhopal, *Amrita Bazar Patrika*, December 16, 1984.

Mitra, Manojit: Smooth start to operation faith, *Amrita Bazar Patrika*, December 17, 1984.

Mitra, Manojit: Unforgettable faces of Bhopal, *Amrita Bazar Patrika*, December 26, 1984.

Mohan Dinesh: Accident or criminal Neglect, *Indian Express*, December 10, 1984.

Mohan Dinesh: Calendar of Negligence. *Indian Express*, February 2, 1985.

Mohan Ram: Counting the cost, *Far Eastern Economic Review* 126(51) December 20, 1984:10-11.

Moncrieff, R.W.: Chemical senses, London, Leonard Hill Books of Morjan-Grampian books, 1967, pp.481-84.

Mukerjee, Dilip: American Response, *Herald Review* 1(14-15), December 9-22, 1984: 15, 17.

Murlidharan, Sukumar: Time to rethink the Quick fix. *Herald Review* 1 (14-15): December 9-22, 1984, 12-13.

Mystery of who designed Bhopal's Plant, *New Scientist*, December 20-27, 1984:7.

Nagpur Times: Criminal Negligence, *Nagpur Times*, December 5, 1985: *Editorial*.

Nagpur times Fleecing the victims, *Nagpur Times*, December 29, 1984, *Editorial*.

Nagpur Times Political Advantage, *Nagpur Times*, December 11, 1984, *Editorial*.

Nagpur Times: Political Advantage, *Nagpur Times*, December 11, 1984, *Editorial*.

Nanda, Meera: Are Pesticides necessary? *Indian Express*, January 17, 1985.

Nanda, Meera: The case for an informed public, *Indian Express*, April 7, 1985.

Nanda, Meera: Killer gas phosgene not MIC, *Indian Express*, December 6, 1984.

Nanda, Meera: Lessons to of Bhopal night none, *Indian Express*, December 26, 1984.

Narain K.V.: Protest in Japan over Bhopal tragedy, *Hindu*, December 21, 1984.

Naqvi, L.H.: Bhopal tragedy, *Tribune*, December 7, 1984.

National Convention on lessons from Bhopal, *Bhopal, February 17-18, 1985*; Declaration, February 18, 1985: Bhopal. Bhopal Zahrili Gas Sanghanrsh Morcha, p. 6.

The operation and after, *Hindu*, December 28, 1984.

No More Bhopal Action, *News Alert*, 2(3), May 1985, 3.

Noorani, A.G.: Bhopal who will pay? *Herald Review*, 1 (16), December 23-29, 1984: 32-33.

Pandhe, M.K.: Bhopal Tragedy who are culprits? *People's Democracy*, December 16, 1984:3,10.

Pandit, Tooshar: Killer Gas callous carbide, *Sunday* 12(7), December 16-22, 1984: 16-25.

Pandit, Tooshar: The great Exodus, *Sunday*, 12(8), December 23-29, 1984: 61-62.

Paranjape, H.K.: The Hypocrisy of Government, *Sunday Observer*, December 16, 1984.

Parigi, Vasant Kumar: Bhopal Massacre: *Consumer Network*, 1(2), January-March, 1985:1. *See also* pp. 4-15.

Parimoo, J.N.: More deathes likely in Bhopal: US experts. *Times of India*, December 7, 1984.

Parimoo, J.N.: Union Carbide stock tumbles in US Market. *Times of India*, December 8, 1984.

Parisar: Lessions of Bhopal: An Environmental statement by parisar, *Poona*, *Parisar*, March 12, 1985, p. 7.

Patriot: Bhopal Herror, *Patriot*, December 17, 1985, *Editorial*.

Patriot: Industrial safety, *Patriot*. February 4, 1985, *Editorial*.

Patriot: Real culprits, *Patriot*, April 3, 1985.

Patwardhan, M.S.: Some major Issues, *Hindustan Times*, January 24, 1985.

People's Democracy: Guility, *People's Democracy* 8(51), December 16, 1984:1, 11.

Pesticides, Perils People, *Amrit Bazar Patrika*, December 6, 1984.

Poisonous gas kills 350 in Bhopal, *Economic Times*, December 4, 1984.

Poser on legal remedy, *Statesman*, December 7, 1984.

Precautions against pesticide Perils, *Hindu*, December 16, 1984.

Press conference, Nirobi, *March 12, 1985*: Press release.

Probe Panel to examine safety, *Patriot*, December 6, 1984.

PUCL Bulletin, People's right to know, *PUCL Bulletin*, March, 1985 3, *Editorial*.

Punwani, Jyoti: Between the coughing the recall Indira Gandhi. *Sunday Observer*, January 26, 1985.

Punwani, Jyoti: Who should victims Sue? *Sunday Observer*, December 16, 1984:7.

Qureshy, Ahtesham: Death dance in Bhopal, *Week end Review*, 5, (42), December 23-29, 84:3. *See also* pp. 4-7.

Qureshy, Ahtesham: Did Buch OK carbide site? *Hindustan Times*, February 13, 1985.

Qureshy, Ahtesham: Poison Gas toll 1200 and Rising, *Hindustan Times*, December 5, 1984.

Qureshy, Ahtesham: US lawyers in Promise dream amounts. *Hindustan Times*, December 31, 1984.

Rahul Dev.: The way the killer Works, *Week* 2(52), December 16-22, 1984: 20-21.

Rajan, N. and Dayal, John: US team barred entry to plant, *Patriot*, December 7, 1984.

Rajimwale, D.A.: Bhopal tragedy, *Indian Nation*, December 10, 1984.

Ramachandran, R.: How could it have accurred? *Frontline* 1(2), December 15, 1984: 66-68.

Ramakrishna, K.: Legal Harles, *Hindustan Times*, December 21, 1984.

Sharma, Vichitra: Bhopal tragedy raises doubts, *Hindustan Times*, December 20, 1984.

Rama Rao, R.: Environment: Problems of developed and developing countries, New Delhi, Economic and Scientific Research Foundation, p. 50.

Ramaseshan, Radhika: Government responsibility for Bhopal Gas Tragedy, *Economic and Political Weekly* 19(50), December 15, 1984: 2109-10.

Ramaseshan Radhika: Profit against safety, *Economic and Political Weekly* 19(51-52), December 22, 29, 1984: 2147-50.

Ramaseshan, Radhika: The man who foresaw the Bhopal tragedy, *Sunday Observer*, December 16, 1984: 7.

Rashtriya Abhiyan Samiti Letter to Prime Minister of India on Bhopal Poisonous Gas disaster, *June 5, 1985*: For Recommendations. *See also* p. 3.

Rathee, K.V.: Bhopal Tragedy: Punitive damages, *National Herald*, January 2, 1985.

Reinhold, Robert: Bhopal plant user up last of chemical, *New York Times*, December 23, 1984.

Reinhold, Robert: Mood of resentment spreads out of Bhopal, *New York Times*, December 16, 1984.

Report on Bhopal slums for U.N. Panel, *Hindustan Times*, February 4, 1985.

Resencranz, Armin: How union carbide can help Bhopal victims, *Statesman*, March 2, 1985.

Robbins, William: Killer carbide prove to kill more? *Amrita Bazar Patrika*, December 26, 1984.

Rye, William A.: Human responses to isocyanate exposure, *Journal of occupational medicine* 15(3), March, 1973: 306-7.

Sadagopal, Anil. Bhopal: Un-answered question, *PUCL* Bulletin, March 1985: 27-28.

Same Danger, *India Today* 10(2), January 31, 1985: 104.

Schaeffer, Robert: Bhopal spurs pesticide reform movement, *Not manapart* 15(6), July-August, 1985: 10-11.

Science Congress to discuss Bhopal leak, *Hindustan Times*, January 2, 1985.

Sensible and safe use of technology, *Hindu*, January 16, 1985.

Shankar, Sandeep: Bhopal mein vayu pralaya, *Ravivar* (Hindi), 8(12), December 16-22, 1984: 42-47.

Sharma, Kalpana: Victims cry out for relief, *Indian Express*, February 3, 1985.

Sharma, Surya Narayana: Massacre at Bhopal, *Commerce*, 149(3837), December 15, 1984: 912-917.

Sharma, Vinod and Singh, G.K.: Tall claims, dubious practices, *Week*, 2(52), December 16-20, 1984: 22-24.

Sharma, Vinod: The Second Scare, *Week*, 3, (1), December 23-29, 1984: 10-11-13.

Shelvankar, K.S.: An issue focus, *Frontline* 1(2), December 15-28: 1984, 68-69.

Shridhar, Vijay Datta and Shrivastava, Mahesh: Bhopal gas disaster and Rehabilitation, *Blitz*, August 17, 1985, Independence day special, 42-43.

Shreff, Rajju: Lessons from Bhopal, *Science Age*, 3(1), January 1985: 27-28.

Shrotriya, Prabhakar: Sun Re Bhopal, Kya Hai Hal Chal? *Sarika* (Hindi), 25(378), 1-15 April 1985: 26-29. *See also* 29-30.

Singh, N.K,: Bhopal six months after, *Indian Express*, June 2, 1985.

Singh N.K.: Gas toll amounts to 1100, *Indian Express*, December 5, 1984.

Singh, N.K.: UCC chief arrested, released on bail, *Indian Express*, December 8, 1984.

Singh, Raminder: Tragic force, *India Today* 10(1), January 1-15, 1985: 103.

Sinha Arti: Murder from the Bhopal gas plant, *Patriot*, December 24, 1984.

Sitomer, Curtis J.: Lawyer ethics is on trial in profile-Union Carbide suit. *Christian Science Monitor*, April 26, 1985: 29-30.

S.K. *Pseud*: How the Gas might have leaked, *Science Age*, 3(1), January 1985: 29-30.

Specious sabotage theory, *Hindu*, May 11, 1985, May 13, 1985.

Statesman, Bhopal's ills, *Statesman* March 28, 1985, *Editorial.*

Statesman: Consigned to limbo? *Satesman* January 15, 1985, *Editorial.*

Hazards of Growth, *Statesman*, December 28, 1984.

Statesman: Playing with poison, *Statesman*, December 5, 1984, *Editorial.*

Statesman: Ritural on safety, *Statesman*, February 7, 1985, *Editorial.*

Statesman: The killer gas, *Statesman*, December 14, 1984, *Editorial.*

Stevens, William K.: Toxic leak: Limited Political effects, *New York Times*, December 23, 1984.

Subramaniam Arun: Catastrophe at Bhopal, *Business India* (77), December 17-30, 1984: 67-70.

Sunday Observer: Shame in Bhopal, *Sunday Observer*, December 9, 1984: 12.

The deadline, *Probe India*, 6(11), January 1985: 36-37.

The Inside story of Bhopal massacre, *People's democracy* 8(535), December 30, 1984: 3-4. *See also* 2-7.

The Killer of Bhopal, *Sunday* 12(10), January 6-19, 1985: 15, Opinions.

Thomas, Vinod: Pollution control in Sauo Paulo, Brazil: Costs, Benefits and effects on Industrial location. *Washington, World Bank*, 1981. World Bank staff working paper, November 1981, no. 501.

Tikku, M.K.: Gas impact study may a few years, *Hindustan Times*, February 17, 1985.

Times of India: Full compensation, *Times of India*, December 7, 1984, *Editorial.*

Times of India: Inexcusable Bungle, *Times of India*, December 10, 1984, *Editorial.*

Times of India, Not by appeals alone, *Times of India*, January 4, 1985, *Editorial.*

Times of India: Search for a scapegoat, *Times of India*, December 13, 1984, *Editorial.*

Times of India: Take over not issue. *Times of India*, January 23, 1985.

Times of India: Tragedy at Bhopal, *Times of India*, December 5, 1984, *Editorial.*

Times of India: Why all this secrecy. *Times of India*, January 30, 1985.

Times of India: Wrongfull confinement, *Times of India*, December 14, 1984, *Editorial.*

Top carbide grass hold Anderson released, *Times of India*, December 8, 1984.

Tribune: Bhopal's black monday, *Tribune*, December 5, 1984, *Editorial.*

Tribune: Bhopal scare continues, *Tribune*, December 15, 1984, *Editorial.*

Tribune: Bhopal the task ahead, *Tribune*, December 21, 1984, *Editorial.*

Tribune: Lessons from Bhopal, *Tribune*, December 6, 1984, *Editorial.*

Tribune: Sinister research, *Tribune*, December 27, 1984.

Trouble on the tarmac, *Probe India* 6(11), January 1985: 34-35.

UNEP: United Nations Environmental Programme, 1979. Nairobi, UNEP, 1779, p. 66.

Union Carbide admits Problems were known. *Guardian Weekly*, 131(25), December 16, 1984: 1.

USA Highlander Research and Education Centre, *etc.* No Place to run. New Delhi, Priya Publications, 1985, p. 39.

Vadlamudi, V.S.V. and Shenai, V.A.: Human responses to isocyanate Exposure, *Science Age*, 3(1), January, 1986: 22-26.

Verma, Dinesh Chandra: Bhopal Gas Durghatana, Pralaya sa Drishya, *Bhoobharati* (Hindi) (178), December (II), 1984: 60-64, 74-75.

Vested Interest in delay, *Statesman*, January 14, 1985. January 15, 1985, *Bhopal Survivors.*

Victims version, *Nagpur Times*, December 6, 1984.

Vikas Ki Neeti Par Ungali Uthao, *Sarvodaya Jagat*, 7(48), December 21-27, 1984: 2. *Editorial.*

Will compansation be clusive? *Nagpur Times*, December 16, 1984.

Witness to a leaking disaster II, *Amrita Bazar Patrika*, December 16, 1984.

Yadav, M.H.: Report submitted to Union Cabinet, *Statesman*, December 9, 1984.

Zahrili gas Kand Sangharsh Morcha Bhopal; Medical relief and research in Bhopal. Report of the All India convension, Bhopal, *February 17-18*, 1985.

INDEX